POLLUTION PREVENTION SOFTWARE SYSTEMS HANDBOOK

POLLUTION PREVENTION SOFTWARE SYSTEMS HANDBOOK

by

Michael F. Wood
John A. King
Nicholas P. Cheremisinoff, Ph.D.

np NOYES PUBLICATIONS
Westwood, New Jersey, U.S.A.

Library of Congress Catalog Card Number: 96-36451
ISBN: 0-8155-1405-0
Printed in the United States

Published in the United States of America by
Noyes Publications
Fairview Avenue, Westwood, New Jersey 07675

10 9 8 7 6 5 4 3 2 1

Library of Congress Cataloging-in-Publication Data

Pollution Prevention Software Systems Handbook by Michael F. Wood, John A. King, and Nicholas P. Cheremisinoff.
p. cm.
Includes index.
ISBN 0-8155-1405-0
1. Pollution--Computer programs. I. Wood, Michael F., King, John A. and Cheremisinoff, Nicholas P.
TD177.P638 1966
363,737'0285'5369--dc20

ABOUT THE EDITORS

Michael F. Wood is an environmental manager with 20 years experience designing pollution prevention, multi-media pollution control and regulatory compliance programs. He holds a B.S. in Environmental Science and a M.S. in Environmental Management from the University of Maryland.

John A. King is president of King & Associates, Inc., and engineering and environmental management consulting firm and is associate professor of Environmental Management at the University of Maryland. He holds a B.Ch.E. from Brooklyn Polytechnical University, and an M.S. from Meridian University. He is the past president and founder of the Hazardous Waste Treatment Council, and the Environmental Business Council, and is a member of AICHE Mr. King is the author of numerous articles and books.

Nicholas P. Cheremisinoff is director of the Environmental Policy and Technology Program on Industrial Waste Management in Ukraine, a project funded by the United States Agency for International Development. Dr. Cheremisinoff has twenty years of applied research and industry experience in the petrochemical industries. He has contributed extensively to the industrial press and has authored, co-authored or edited over 140 reference books, including Noyes Publishers *Hazardous Materials and Waste Management* and *Transportation of Hazardous Materials*. Dr. Cheremisinoff received his B.S., M.S. and Ph.D. degrees in chemical engineering from Clarkson College of Technology. He is a member of Tau Beta Pi, Sigma Xi and Omega Chi Epsilon.

DISCLAIMER

This book does not imply or explicitly endorses the use of any specific software system listed. System profile information was collected from a variety of sources including DOE reports, research reports for Federal agencies, promotional literature written by tool producers, and conversations with tool producers. In addition, system producers received copies of these system profiles and reviewed them for accurac, and their comments were incorporated into the survey.

ACKNOWLEDGEMENTS

Most of the data and information describing pollution prevention and other chemical information management systems was collected by Holly Ellwood of the Pollution Prevention Division of the U.S. Environmental Prevention Agency and reviewed by the following team of reviewers:

Claire Barth - Senior Editor
Institute of Management Accountants

Louis Bisgay
Director, Management Accounting Practices
Institute of Management Accountants

Robert Bollinger
Health, Safety, and Environmental Consultant
Environmental Health and Safety (EH&S)
Software Development Group

Don Brennan
Environmental Information Manager
Environmental Regulatory Affairs Department
Rohm and Haas

Rick Brenner
Planning, Prevention, and Compliance Staff
Federal Facilities Enforcement Office (OFFE)
USEPA

Greg Daro
Ecology Information Team Leader
Ecology and Safety
BASF Corporation

Marc Epstein
Visiting Professor
Stanford University, Graduate School of Business

Julian Freedman
Director of Research
Institute of Management Accountants

Eun-Sook Goidel
Pollution Prevention Division
Office of Pollution Prevention and Toxics, USEPA

Bill Hombach
Technical Director
Communications Training Analysis
Corporation (CTAC)

Mike Kabjain
Project Manager
EMAX Solution Partners

John King
Senior Regulatory Consultant
Resource Consultants Inc.

Gerald Kos
Chief, Cost Estimating Branch
EER Systems

Timothy C. McManus
Senior Principal
Project Management Associates, Inc.
Chairperson, Government Liaison Committee
AACE International

Marty Spitzer
Pollution Prevention Policy Staff
Office of the Administrator
USEPA

Peter Thompson
Pollution Prevention Division
Office of Pollution Prevention and Toxics
USEPA

Jeannie Wood
Manager
AT&T Global Information Solutions

PREFACE

The *Pollution Prevention Software Systems Handbook* is an up to date review of major software products that are available for addressing waste management and pollution prevention opportunities. In recent years there has been an explosion in software systems applicable to the environmental consulting and management fields. The chemical industry early on recognized the value of the computer in process design and control applications, and particularly in project management for large scale operations. Environmental engineers, consultants and environmental managers have only come lately to rely and, indeed, demand more sophisticated software systems to address waste management and large scale environmental management and project issues. Approximately twelve years ago one of the editors of this work was asked by *Pollution Engineering Magazine* to review environmental software products. At that time, there were only a few dozen such products, despite enormously high levels of environmental remediation projects among industry players. Today there are hundreds of such programs with a wide range of sophistication, prices and specialized applications.

This handbook provides a review of what the editor's consider major software that emphasize pollution prevention applications. Neither the editors nor publisher imply endorsement of any software system described, nor should the exclusion of any product from this handbook be construed as a negative aspect of any product. A special thank you is extend to Mr. Pavel Nagopetian for assisting in the preparation of this book, and to Noyes' Publishers for their fine production.

Nicholas P. Cheremisinoff, Ph.D.

CONTENTS

ABOUT THE EDITORS v

DISCLAIMER vi

ACKNOWLEGEMENTS vii

PREFACE ix

CHAPTER 1. INTRODUCTION 1
CHEMICAL TRACKING 2
HISTORICAL PERSPECTIVE 4
SYSTEM SELECTION CRITERIA 13
EIGHT CLASSES OF SYSTEMS 17

CHAPTER 2. SYSTEMS OVERVIEWS 23
COST-ESTIMATING SOFTWARE SYSTEMS 23
LIFE-CYCLE ANALYSIS (LCA) 24
COST ESTIMATING SOFTWARE 25
RISK ASSESSMENT SYSTEMS 34
REMEDIATION SYSTEMS 37
FINANCIAL ANALYSIS SYSTEMS 44
ENVIRONMENTAL COST IMPACT ANALYSIS 47
WASTE REDUCTION SYSTEMS 50

CHAPTER 3. COST ESTIMATING SOFTWARE 53
ComposerGold/MCACES 53
COSTPRO 61
Freiman Analysis of System Techniques (FAST) 68
G2 Estimator 76
Hard Dollar EOS 84
Precision Estimating Plus 93
Success 99
TRACES 106

ABBREVIATED PROFILES OF COST ESTIMATING 112
ASAPMS Housekeeper 112
Cost Engineer 113
Cost Estimating 115
Enhanced Cost Estimating Relationship Program 115
Esceomate Piping/DWV 116
PACES (Parametric Cost Engineering System) 117
Pulsar 118
Questimate 8.1 120
Richardson RACE (Rapid Access Cost Estimating) .. 122
StreetSmarts 123
WinEst 124

CHAPTER 4. SCHEDULING AND COST CONTROL .. 127
HIGH-END SCHEDULING 127
Dekker TRAKKER™ Project Management System ... 127
Open Plan Professional 132
Primavera Project Planner (P3) 138
Project Execution System (PXS) 143
SAS/OR 148
Project Scheduler 6 Version 1.5 156
Texim Project for Windows 161
COST/CONTROL ANALYSIS 166
Cobra 2.0 166
Cost Time Management (CTM) 172
EasyABC Plus 177
Parade 183
ABBREVIATED PROFILES 188
FOREFRONT 188
P/X 189

CHAPTER 5. RISK ASSESSMENT SYSTEMS 193
GENERIC RISK ASSESSMENT 193
MOUSE (Modular Oriented Uncertainty SystEm) 193
Project Risk/Contingency Independent Cost

Estimating Contingency Analyzer (ICECAN) 198
Monte Carlo 202
Range Estimating (REP) 206
ABBREVIATED PROFILES 211
Environmental Risk Management Information System (ERMIS) 211

CHAPTER 6. REMEDIATION SYSTEMS 215
COST ESTIMATION AND SCHEDULING SOFTWARE . . . 215
HAZRISK (database and models) 215
Remedial Action Cost Engineering and Requirements System (RACER)/ENVEST 219
Remedial Action Cut Estimating System) (RACES) . . 226
RI/FS: Planning and Risk Analysis RAAS , MEPAS and ReOpt 230
COST TRACKING 236
Removal Cost Management System (RCMS) 236

CHAPTER 7. TRACKING SYSTEMS 243
FACILITY MANAGEMENT AND HEALTH & SAFETY . . . 243
AWARE 243
DataPipe 250
Environmental Managmmnt System (EMS) 257
Facility Tracking System (FacTS) 265
FLOW GEMINI Information System-EIS, OHIS 272
TINIA 278
WASTE MANAGEMENT AND TRACKING 285
Cradle to Grave Version 2.0 285
Docuwaste, CHEMTOX Systems 291
ENFLEX Data 299
G2 Real Time Expert System, GOA 306
Gentrax, and Other Wixel Products. 313
HAZE 321
HAZMIN, and other LOGICAL products 328
Material Inventory Report System 336

PlantWare 342
WASTE DB TM, and Other Envirogenics Products . . . 349
Waste Trail and other Corbus Systems 355
ABBREVIATED PROFILES 362
Software from Environmental Software Systems Inc. . 362
Spreadsheet Templates (MS-DOS and Macintosh) . . . 364
Programs Developed for/by the USEPA 364
Other Software 365

CHAPTER 8. FINANCIAL ANALYSIS SYSTEMS 369
Costing, Financial Analysis of Pollution Prevention Projects . . 369
Economic Analysis for Pollution Prevention 373
P2/FINANCE: Pollution Prevention Financial Analysis - Cost Evaluation System 378
Pollution Prevention Benefits Manual 384
PRECOSIS 389
ABBREVIATED PROFILES 394
PAPA Investment Model 394

CHAPTER 9. ENVIRONMENTAL COSTING IMPACT ANALYSIS SYSTEMS 397
APACS Life Cycle Cost Software Demonstration 397
ECO-SYS 401
EE Energy/Environmental Life-Cycle Assessment 408
The EPS Enviro-Accounting Method 413
Hazardous Material Life-cycle Cost Estimator (HAZMAT) . . 417
Hazardous Material Life-Cycle Cost Model 425
Life Cycle Costing Program-Version 2.0 429
Total System Life-Cycle Cost Analysis 436
Tools for Environmental Analysis and Management 440
Tellus Packaging Study 444
ABBREVIATED PROFILES 449
Abridged LCA Matrix 449
ASEAM (A Simplified Energy Analysis Method) 451
BLCC (Building Life Cycle Cost Program) 452

FEDS Release 3.0 455

CHAPTER 10. WASTE REDUCTION SYSTEMS 459
Avoidable Waste Management Cost at DOE Facilities 459
SRS Waste Cost Analysis 463
Systems Cost Model (SCM) 468
Waste Cost Avoidance Model Template 474
Waste Life-cycle Cost: Generic Guidance and Assumptions . 478
WastePlan 482

REFERENCES 487
GLOSSARY 495
INDEX 517

CHAPTER 1. INTRODUCTION

This book presents a view of environmental health and safety systems now available for management functions along with simultaneous capabilities to effect pollution prevention (P2). The focus is on systems with pollution prevention potential, along with capabilities relating to incorporating costs into decision making. There is an explosion in the number and type of computer systems available to assist environmental managers with various aspects of P2. It should prove a useflull resource for these managers and other researchers, as it characterizes and lists the many chemical information systems that directly or indirectly facilitate pollution prevention audits, projects and program designs.

Many chemical information management systems have been designed and used to manage chemical information related to environmental health and safety protocols relating to the manufacture, purchase, transportation, storage, use, and disposal of chemicals. Requirements such as the need for readily accessible Manufacturer Safety Data Sheets (MSDS), easily prepared and accurate chemical inventory reports and release data required by the Emergency Planning and Community Right to Know Act (EPCRA), recordkeeping mandated by waste disposal controls imposed by the Resource Recovery and Conservation Act (RCRA) and state waste laws, and complicated hazardous materials transportation requirements mandated by the Department of Transportation (DOT) have forced firms to adopt aggressive and sometimes complicated approaches to maintain chemicals and records so as to be in compliance. As chemical management has become more complex and resource intensive, the need for chemical management information systems has been realized. Numerous captive systems have been developed by larger firms and thousands of software packages are commercially available from software developers and environmental consulting firms. The environmental software field is now large enough to support specialized trade shows and conventions.

CHEMICAL TRACKING

The use of computer systems for chemical tracking and pollution prevention and to enhance environmental management and compliance with environmental requirements is an emerging opportunity for software developers and environmental managers. Systems can be used in the chemical manufacturing and other manufacturing industries which process or use chemicals in their business, the waste disposal industry, and government facilities. Chemicals today are pervasively regulated to the point where normal materials used in manufacturing are defined as chemicals under the federal laws.

Chemical manufacturers have been heavily regulated and have, for a number of years, pioneered the use of computer systems for a myriad of uses such as process control, process optimization, and waste minimization. As a result of regulation, public scrutiny and economics, the chemical industry has used computers as a tool for various business purposes including chemical tracking for environmental regulation compliance. Use of computers for chemical tracking and overall chemical management emerges as the initial opportunity for the chemical industry, software firms, and government regulators who must fashion controls that are achievable and economically feasible.

In contrast to the chemical industry, general manufacturing facilities, while heavily dependent on chemicals and chemical compounds which are regulated, have not always been as fast to recognize their environmental issues and implement programs to ensure compliance. Obviously, in certain areas environmental controls have been stringent and long-standing. For instance, steel production and electric generation, have received steady attention and controls have been imposed to reduce air pollution. In another example area, pulp and paper with its heavy reliance on water, has been the focus of water pollution prevention and control requirements. As regulations and economic needs become more pervasive for the overall control of chemicals in all media, and their impacts reach diverse manufacturing sectors as furniture and electronics, these industries are responding with more proactive environmental health and safety (EH&S) programs to ensure compliance, present a "green" image to customers and the public, and effect economic gains.

The waste disposal industry itself presents a third area of opportunities for chemical information management systems. Controls imposed under RCRA for cradle to grave tracking of wastes create circumstances under which systems recordkeeping and tracking functions can be successfully applied. Finally, government organisations and facilities which purchase, store, use, and dispose of chemicals now use systems to enhance their EH&S management capabilities and maintain compliance. Some organisations such as the U. S. Navy, routinely purchase, store, use, and dispose of large quantities of chemicals and chemical products.

On the positive side, potential benefits of the use of systems include improved health and safety, improved compliance with environmental laws and requirements, business improvements such as resource savings and efficiencies from reduced personnel costs and improved time savings, and access to information which could assist with pollution prevention and waste minimization efforts. On the other side of the coin possible negative impacts resulting from the use of systems could include: errors in systems resulting in noncompliance; health, safety, or environmental risks, the potential for government policies producing a drag on development and use incentives, and overly complicated systems not meeting users needs, especially for small businesses or for extremely large applications.

HISTORICAL PERSPECTIVE

This introduction provides a historical background of the use of chemical information systems, that led to today's applications for EH&S (Environmental Health and Safety) and simultaneous pollution prevention initiatives. Software for environmental management emerged as the PC began to filter into industrial and environmental management operations. Kumar and Sahore in (1993) relate that the field of environmental engineering has numerous problems requiring powerful tools for development of solutions. Use of the computer has helped the environmental professional in many ways during the last two decades and spawned development of many useful software packages. Kumar and Sahore (1993) go on to estimate that there are approximately 3,000 environmental software packages available. Esry (1993) notes that differences between packages can include: the computer platform, costs, user interface, standard preprogrammed reports, graphics capability, update and maintenance policy, help screens and documentation, hot line support, data entry and transfer services, future system expansion updates and associated costs, ability to customize or modify the package to suit individual needs, and integration services. The possibility of integration with existing computer systems and processes is increasing in importance. As Luce (1990) concluded that over time, EH&S programs and management have merged into an integrated management approach and systems are being designed which recognize this.

A long term reviewer of environmental software is Gerald Rich. He writes for a number of publications and has been preparing reviews since 1985. The 1992 review in *Pollution Engineering* (Rich, 1992) notes the sophistication of environmental software has increased dramatically since *Pollution Engineering* began to review programs in 1985. The very first review, done by Cheremisinoff and Cheremisinoff covered products from only 20 companies. In 1989 Rich indicated that over the years, each year saw a doubling of the programs available starting with about 25 programs in January 1985, 50 in 1986, 100 in 1987, 200 in 1988, and 400 in 1989. Rich felt the increase was a result of smaller companies providing programs for niche type problems such as dispersion modeling along with the tremendous expansion in the capability of computer systems, allowing far more comprehensive programs to be developed such as for the various RCRA programs encompassing the entire realm of environmental concerns. Rich observes that there was also a doubling in the number of pages of regulations, adding to the complexity of an environmental manager's job during the five years noted above.

Decision-makers in industry and in federal facilities rely on tools and software systems everyday to help them make decisions. McAdams and McAdams (1995) note, however, that despite the proliferation of information technology throughout most facets of American life, and the fact that certain fields such as banking, finance, and accounting have been quick to embrace the new technology, the environmental industry is a latecomer to digital information systems. These authors theorize that the low tech "dirty shirt" roots of the industry foster conservative executives who are reluctant to introduce new technology without well documented evidence of economic benefits. Now that the technology is affordable, portable, and well proven, the environmental industry is moving into this field which has become universal to large colonies. In fact with the work on such standards as ISO 14000 environmental management systems and approaches to activities such as audits are becoming both standardized and international, providing additional impetus for adoption of computer technologies in the environmental field (Kirschner, 1995). This point of view is in agreement with this book's coverage of available tools for environmental management. As noted in this text, organizing for ISO 14000 is creating a need for systems approaches.

Chemical Industry Takes The Lead

The chemical manufacturing industry has been using computer technology extensively for a number of years. In the chemical manufacturing sector, process simulation software is available to optimize yields and *minimize* environmental issues. Sowa (1994) reports that process simulation tools have been used extensively by the petroleum refining and commodity chemical industries for the past 25 years. Commercial products have replaced early in-house process simulation tools over the past ten years. They can significantly improve the performance of batch chemical operations. These systems can answer key questions and provide data to resolve issues of process safety and waste minimization. Sowa discusses issues such as direct solvent recovery to manage solvents at the end of a batch, liquid extraction to remove 'bad actors', vapor/liquid absorbers to control vapor emission and determining whether the vapor above a liquid stored in a tank is flammable. A recent but similar article by Shanley (1995) reviews the various types of software packages available for process simulation and discusses uses for reducing process pollution.

Esry (1993) notes that the Superfund Amendments and Reauthorization Act of 1986 (SARA and EPCRA) changed the nature of environmental data management and compliance reporting for most companies. Esry (1993) believes that the repetitive nature of the reporting and the complicated calculations for the regulation of thousands of chemicals and the products they make up made manual methods unpractical. In addition, severe penalties for reporting errors, late reporting, or failure to report, inadequate documentation or improper records retention make environmental compliance reporting as critical to business survival as cost and tax accounting. This author goes on to highlight that EPCRA chemicals constitute tens of thousands of materials, and that the OSHA Hazardous Communication Standard also requires data management and reporting for most chemicals regulated under SARA and mandates MSDS's. Donley (1991) indicates that a small manufacturer can have a few hundred MSDS's to track, while a large industrial facility could have 10,000. Guggenhiem (1994/1995) highlights that manually assembling the requisite information and performing detailed inventories on a broad range of chemicals and chemical products in geographically and functionally dispersed operations is prohibitively expensive.

Esry (1993) relates that most of the reporting software programs are written as applications around standard commercial database management systems (DBMS) like Oracle, Focus, and Progress, or are written completely in a language like Ñ, COBOL, etc. *Chemical information systems* usually provide a template for data entry, allow required calculations and unit conversions and contain a report generator to provide the user a means of extracting the data directly from the database. For existing chemical systems, Rich (1995) uses 34 categories to describe the software package capabilities. These range from corporate multiplant management, permit tracking, through mapping/GIS to laboratory management. McAdams, and McAdams (1995) list 6 areas where the applications are currently centered. These are: improving waste collection, transport and disposal efficiency; improving efficiency and effectiveness of remediation, comprehensive decision making tools; efforts to make industrial processes cleaner; and source reduction. Donley (1993) categorizes environmental software as covering six categories which are: hazardous substances, water, mapping, air, ground water, and miscellaneous.

Guggenheim (1995) reports on use of systems by large electric utilities to integrate a broad range of chemical information in order to improve regulatory compliance, facilitate chemical management, identify opportunities for waste reduction, and support research applications. The main thrust of the computer systems was to link the existing MSDS database with the procurement database. One advantage was thought to be the cost- effectiveness of eliminating large initial data input requirements with this approach. The utility company in Guggenheim's 1995 review saw a number of objectives for CIMS such as decreasing the amount of time needed for regulatory requirements and research questions, and a cost effective alternative for traditional product inventory.

In *Environmental Protection,* Rich (1995) relates how crunching numbers, inventorying air emissions for possible CAA (Clean Air Act) Title V compliance was one of the latest examples of the need for automation tools. Rich (1995), in discussing air program compliance applications lists 10 categories of computer databases including: permit tracking, process emissions calculations, fugitive emissions calculations, dispersion modeling, and facility management.

Chiu and Shoberg (1995) provide a number of examples of applications. They describe a computer system that ties together different pieces of chemical management into one database to streamline EH&S process review. Once an MSDS is entered into such a system, information on the chemical or product involved can be accessed by multiple users to meet the hazard communication standard. Characteristics of the chemical such as flammability can be used for queries, and information from various lists can provide regulatory compliance data. One example given was acetone, which is on the EPCRA chemical list, and is also listed under the TSCA 12 *(b)* export notification list and as a VOC under the CAA (Chiu & Shoberg, 1995).

Systems are available in various price ranges depending on type of system, extent of modifications for a user's specific needs, system capabilities and complexity, number of users, number of modules purchased, etc. Donley (1994) notes software prices range from no charge to $80,000, with most systems costing less than $3,000. Costs are provided where available in this book.

P2 Value Added

Companies now recognize that initiatives such as proper materials and waste management, efficient product and process design, energy efficiency, and in short, Pollution Prevention (P2) can be both profitable and environmentally compelling (Ditz, Rananathan, and Banks, 1995).

The increasing focus of the public and legislators on simaltaneously optimizing energy use and enhancing the environment has created the interaction of manufacturing operations economics and environmental management (Klassen, 1993). Environmental management has become a strategic business issue and is being integrated into the business processes of design, production, delivery, circulation, use and disposal of products. Klassen (1993) notes that as firms strive to reach the highest level in the hierarchical approach to environmental excellence, environmental, improvement objectives, like quality, must become part of the annual business plan, with performance reviews to track progress. A long term competitive advantage may be possible by moving ahead of the industry, setting industry standards, and influencing customer adoption of sound environmental, products and processes (Klassen and McLaughlin, 1993). The complexity of interactions between environmental issues and manufacturing necessitates an integrated, coordinated effort to respond appropriately.

On a specific firm or facility basis, Heilshom and MacDougall (1992) feel that the decision to use a manual versus a computer system for environmental management can depend on many factors such as company size, number and complexity of waste streams, and other reporting requirements. The complexity of a company's or a plant's waste management process will probably be the main determinant as to whether the company would benefit from computerized systems. One expected efficiency, cited by Rich (1995) when he notes a simplistic assumption regarding computer systems, that work hours can be simply minimized through using computer software. Chic and Shoberg (1995) report on IBM's efforts to enhance their chemical management process to keep pace with regulatory, industry, and company requirements. They note that computer based information systems have played a key role in: tracking process chemical inventory, maintaining EH&S data, MSDS management, waste tracking, and regulation tracking. Randall and Shih (1990) describe General Motor's use of one package (MIRS by AV systems) for SARA Title III, other compliance reporting, and as a tool for managing the facility's chemical inventory data. The system reliably produced all required reports. Add hoc report generation for management surveys proved valuable as well. In monthly meetings of the in-house Hazardous Materials Committee, CIMS generated reports provided a means for communicating the participating departments' hazardous chemical information. This included safety and health, waste treatment, engineering, security, production, material, traffic, and purchasing departments all working on different aspects of hazardous chemicals with printouts keeping information in focus. The system has also proved useful in providing credibility outside the facility.

Experts in pollution prevention such as Chada (1994) note that computerized data bases used to track emissions and wastes can be useful tools for identifying and targeting key sources for pollution prevention. Manufacturing facilities must deal with thousands of chemicals, mixtures and commercial products in their inventories, and reports must assemble data from each MSDS and many other sources to get the information required for each of these reports. In addition, each plant must be subdivided into logical reporting areas and the quantity of each material present must be determined (Esry, 1993). The General Electric Co. (1993) has a program for improving chemical management which stresses the integration of chemical management responsibilities across all organizational departments. Their approach manages design, ordering, receiving, storage, distribution, and use, recycling, and waste accumulation and disposal. One portion of this overall effort includes use of software tools to assist in the compliance effort. More importantly, this system provides a three pronged rational for chemical management. These three reasons are: 1) compliance with legal requirements, 2) ethics, doing the right thing, and 3) good business, productivity and cost. The goal is 100% compliance which includes: no violation of the 90 day rule, no missing MSDS, no lack of employee Hazard Training, no unpermitted discharges, no improper storage, and no missing right-to-know reports.

To date, the major focus of many EH&S professionals has been on cost and compliance rather than on strategic business advantages that could be derived from a "paradigm shift", e.g., a new way of looking at EH&S opportunities. EH&S risk should be viewed on a par with other strategic issues such as financial risk. EH&S risks include worker health and safety, public health and safety, natural resources, public goodwill, as well as financial assets. (KoUuru.1995).

Hammer and Champy (1993) define reengineering as “the fundamental rethinking and radical redesign of business processes to achieve dramatic improvements in critical contemporary measures such as cost, quality, service, and speed”. They indicate that in reengineering, information technology (IT) is an essential enabler, without which, processes cannot be reengineered. A premise of Hammer and Champy (1993) is that companies need to make technology exploitation one of their core competencies, if they are to succeed in a period of ongoing technological change.

Davenport (1993) in his work on reengineering notes that by virtue of its power and popularity, no single business resource is better positioned than IT to bring about radical improvement in business processes. Davenport (1993) reviews the fact that it may take many years for companies to fully incorporate a technological advance within their organization. He feels that IT at a macro level may have fallen short of its promise for effecting business transformation, and stresses the need for IT investment linked specifically to process change.

Lorenzi (1994) notes that while annual financial reports are a staple of corporate America today and he predicts that someday annual EH&S reporting will be as common as annual financial reporting. He relates that being green is no longer synonymous with losing money. Lorenzi stresses the need to manage our destiny before it manages us. Lorenzi (1994) indicates that that EH&S will inevitably be part of the next wave of re-engineering. He views the key as understanding EH&S functional management processes, and their linkage to broader processes critical to managing a business. Lorenzi (1994) stresses linking changes in EH&S to product/process development, equipment, design/development, manufacturing, purchasing, and logistics decisions. Reengineering integrating EH&S offers a way to improve performance on multiple fronts simultaneously (Lorenzi, 1994). In agreement with the IBM and GE approaches described earlier, Lorenzi (1994) notes that the key is to integrate responsible environmental management into all business units. Guggenheim (1995) in his research notes important CIMS uses resulting to regulatory requirements including: employee right to know issues, community right to know, estimation of air emissions (e.g., California AB 2588 - Toxic Hot Spots and SARA section 313 TRIO, and DOT requirements for transportation. Additional important applications relating to tracking information about chemical use include: controlling procurement especially by restricting purchases to only company approved materials, understanding consumption patterns, and industrial hygiene/epidemiological applications. Guggenheim's 1995 article listed a number of objectives for environmental computer systems, such as the decrease in the amount of time needed for regulatory requirements and research questions, and cost effective alternative for traditional product inventory.

SYSTEM SELECTION CRITERIA

Chiu and Shoberg (1995) indicate the most important factor in defining a system is to understand the activity or business process that CIMS will support. IBM developed their Chemical Health and Safety Environmental Management System (CHEMS) after their EH&S process assessment. This view is bolstered by McLaughlin's (1995) assessment that in selecting software, you must begin with a compelling business case that establishes the business imperatives and outlines estimated costs and expected benefits.

Additional criteria laid out by Chiu and Shoberg (1995) for system design and function include: providing for maintenance, distribution, and shared access of chemical information to all users involved in the company's chemical management process. They stress use of a relational database and implementation on a local area network (LAN). The relational database rather than text technology helps support ad hoc searches which are more powerful than key word searches.

Rich & Shih (1991) review a number of selection factors for software including large versus small vendors, modular flexibility, price versus value, and EPA approval. In the approval area EPA had preapproved formats for magnetic data submissions of TRI reports to ensure reliability and accurate data entry. The ERA reviews vendors applications and approves formats that match the approved Form R Toxic Chemical Release Inventory Reporting Form and the EPA Automated Form A or AFR (EPA, March 1995 and EPA, June

1995). These authors list a series of 31 questions to ask before you *buy* including: can the program interface with other computers?, and can you legally back up the program? Rich (1992) reminds that regulations are continually changing and systems selected must handle regular updates.

Wu (1990) notes that there are no precise guidelines for selection of application packages. Formal procedures designed solely for ensuring the proper fit of the application package to the organization's problem do not ease in the same way that the use of structured methodology has been established for analysis and design of a new system. Wu (1990) lists a number of recommendations, and cautions that packages must be studied in more than a cursory fashion. She also advises that subsequent modification of packages may not be successful, and one should check with other users of packages regarding their experiences with vendors and software prior to purchase.

Esry (1993) recommends first determining functional requirements by conducting a needs assessment, then evaluating about 20 packages ensuring to involve the users in the process. Saarinen and Vepsalainen (1994) relate that more and complicated standard requirements indicate the use of outside consultants or software contractors who have experience and knowledge about similar types of systems. These authors see that for routine systems common in many organizations, acquisition and tailoring of a software package provides the most efficient procurement strategy. In discussing choosing software for wharehouse control systems which can be related to chemical tracking, Nixon (1994) notes a number of steps and criteria. He stresses defining needs as a first step prior to selecting a software vendor. Typical selection criteria this author lists include cost, system design, operational design, training and documentation, and support capability.

There are numerous laws and regulations at the federal, state, and local level which impact both the need for and type of computerized information a facility or organization uses in environmental and health and safety management. This spans both traditional federal environmental statutes such as RCRA, and EPCRA, federal requirements impacting transportation such as the DOT rules (49 CFR 107, et al.), and the Hazard Communication Standard adopted by OSHA (29 CFR 1910.1200) along with other OSHA worker health and safety requirements.

Sheng, Dornfield, and Worhach (1995) raise the issue of how government policies can facilitate the implementation of green practices. They view two components as important: (1) creating an awareness among engineers of environmental issues to aid the level of consciousness in products and processes, and (2) developing an understanding of the linkages between operation and manufacturing activities and the adoption of green practices among corporate management and government policy-makers. In their view, management can create strategies that will facilitate rapid adoption, and government can create policy incentives that reflect capabilities and limitations of organizations.

The authors have included an assessment of how three EPA policies that could impact on the enforcement of environmental standards, and also have a positive or negative affect on an organization's use of CIMS. These are briefly reviewed below. The voluntary environmental self-policing and self-disclosure policy (audit policy) is a policy EPA has announced that provides certain incentives for facilities which conduct voluntary compliance evaluations, including audits, and disclose and correct any violations detected. EPA will not request access to evaluation documents to trigger enforcement actions, and will substantially reduce any gravity based penalties associated with any civil matters and will not refer cases for criminal prosecution where certain conditions are met. Various conditions must be met to meet the provisions of the policy including: discovery of violations through voluntary means, disclosure of violations, and prompt correction of problems. This policy is designed to reward environmentally responsible actions and encourages self-monitoring and self-correction of violations (Herman, 1995). Use of computerized records and systems may assist industry in environmental reviews and audits to not only maintain compliance, but quickly identify and remediate situations of non-compliance.

Another EPA policy, the Supplemental Environmental Management Projects (SEP) Policy refers to an EPA approved policy of promoting environmentally beneficial projects as part of settlements in environmental enforcement actions. Under the SEP policy, firms agree to undertake certain projects, and some percentage of the cost of the project is considered a factor in establishing the final penalty to be paid (EPA SEP Fact Sheet, 1995). These projects may fall into any of seven categories such as: pollution prevention activities to reduce the generation of pollution through source reduction, and assessments and audits which include environmental management and compliance assessments and audits (*60* Federal Register, 1995). Systems could in some circumstances assist a facility in implementing or carrying out a SEP. Again, the policy is designed to allow for flexible and common sense settlement of enforcement actions (EPA-May 4, 1995). Included in SEP projects are pollution prevention efforts. An excellent recent treatment of this subject keying on specific enforcement policies is presented by Seeker and Ashford(1995).

In a January 9,1995, memorandum to EPA Administrator Carol Browner, EPA Assistant Administrators Lynn Goldman and Jonathan Cannon set out a proposal for one stop public access and reporting (consolidated reporting). This EPA proposal for a new information collection/management initiative addresses the problem of various (and different) collections of environmental data required under a variety of federal and state statutory and regulatory authorities. The current system is seen as burdensome to industry and makes data use by regulators, business, and the public difficult. The EPA proposals relate to establishing a consolidated (one-stop) approach to collection of data. Environmental information collections which obtain information concerning environmental releases, transfers, emissions, and permits generally would be consolidated into a single, once a year, report to EPA. The impact of such a proposal may include reducing the reporting burden on industry and could obviate the need for chemical information management systems.

EIGHT CLASSES OF SYSTEMS

Decision-makers in industry and in Federal facilities rely on tools and software systems everyday to help them make decisions. When these decision-support products incorporate environmental costs and considerations, they can save companies and Federal facilities money while simultaneously protecting the environment. However, if environmental cost information remains hidden and unaccounted for in the decision-making process, well-informed decisions on environmental management and investments cannot be made. Decision-support products that do not consider environmental costs and considerations may hinder optimal decisions. Evaluating the extent to which existing decision-support products include environmental information is one important task towards incorporating environmental costs and considerations into business and Federal facility decision-making.

Companies and Federal facilities currently invest significant amounts of time and money to comply with growing environmental regulations. Environmental spending is now approaching 2 percent of the nation's gross national product. U.S. companies spend an estimated $140 billion annually to improve environmental performance, while only a quarter of these costs are traced directly to environmental compliance (Owen, 1995). Better management of environmental costs is important for reducing public and private expenditures and for avoiding contingencies in the form of liabilities. Companies and facilities also recognize that initiatives such as proper materials and waste management, efficient process and product design, energy efficiency, and recycling can be both profitable and environmentally preferable (Ditz, 1995).

In addition, new standards and mandates are encouraging companies and Federal facilities to better manage their environmental costs and considerations. International standards are now requiring companies to develop environmental management systems. For example, the International Organization for Standardization (ISO) is creating a series of voluntary environmental management standards entitled ISO 14000. Conformance with standard ISO 14001, the Environmental Management System (EMS) specification document, has strong potential for becoming a de-facto requirement for conducting trade, both domestically and internationally. Procedures and tools that carefully track environmental costs and impacts will greatly assist organizations in meeting the monitoring and performance measurement requirements of ISO 14000. Another change being seen in the private sector is that consumers are beginning to ask product suppliers to take back products when they can no longer be used, or are asking for information on the environmental impact of a product prior to purchase. Federal facilities are now required, under Executive Order 12856* to apply life-cycle analysis (LCA) and total cost accounting (TCA) principles to the greatest extent practicable when estimating pollution prevention opportunities. In addition, Executive Order 12873 requires Federal facilities to give preference to purchasing environmentally preferable products.

The software and systems are categorized into the following eight classes, each covering a particular aspect of management or environmental management:

Class I-cost-estimating software systems: products designed for cost estimation. These systems are usually used by project managers to estimate conceptual and/or detailed project costs.

Class 2-scheduling and cost control/analysis software systems: different products designed to automate project scheduling, monitoring, cost control, and performance measurement. This also includes a couple of Activity Based Costing (ABC) systems. Such systems are used by project managers to maintain project schedules and budgets according to plan. The systems under this section are subdivided in four categories: a) High-end scheduling b) Low-high end scheduling c) Low-end scheduling d) Cost control/analysis.

The subdivisions for the scheduling systems do not have any hard and fast definitions. The terminology is based on language used by vendors and how they refer to the scope of their products. Scheduling systems for performing relatively simple jobs with the minimum required features (e.g., start and finish dates and critical path) are loosely called systems for low-end scheduling. Similarly, low-high and high-end scheduling systems indicate a higher level of sophistication of the system and a greater number of features.

Class 3-risk assessment and contingency analysis software systems: products designed to assess risk (not necessarily project related), including some systems that are specifically designed to compute project-related risks and contingency. The software systems covered under this section have been further subdivided into two subcategories: a) Generic risk assessment b) Project risk and contingency.

Generic risk assessment covers software systems that compute statistical uncertainties associated with static mathematical models defined for various areas (e.g., environmental impacts, costs, etc.).

On the other hand, the systems in the second category have been tailored to address specific areas. Monte Carlo simulation is an approach that many risk analysis tools employ, although one of the tools covered has "Monte Carlo" as its trade name.

Class *4* -environmental management and regulatory compliance software systems: products designed to track waste and waste management costs, primarily to comply with various environmental regulations. Some software systems covered under the section on environmental management and regulatory compliance have features for project management or the potential to include such features. The software systems covered under this section have been further subdivided into two categories: a) Facility management and environmental health and safety b) Waste management and tracking.

Class 5-remediation project-related software systems: software products designed to address different areas of remediation projects. The tools covered have been further subdivided into three categories: a) Cost estimation and scheduling b) RI/FS planning and risk analysis c) Cost tracking

Class *6* -financial analysis software systems and tools: software systems or tools designed to provide information and methods for the financial analysis of projects. These primarily address, but are not limited to, pollution prevention (P2) projects. These tools can be used by managers to calculate and compare financial returns from projects through the use of different indicators.

Class 7-environmental LCC and impact analysis software systems and tools: a broad range of software systems or tools that identify, estimate, and/or analyze environmental costs and impacts for specific projects, areas, or as generic tools. These tools can be used by environmental managers as methods of analyzing and managing environmental costs.

Class *8* -waste reduction software systems and tools: software systems or tools designed to provide general or project/facility-specific guidance on methods to effectively manage wastes or environmental costs. These products only indirectly address compliance issues [This section includes a subsection on proprietary software systems and tools]. Also included is a section on case studies, which depicts the application of particular methodologies. These products address various approaches and methods of environmental management and can be used by business and facility managers.

These eight classes cover the management and decision-support products that play important roles in the business and Federal facility decision-making and management process. Many of the products in Classes 6, 7, and 8 are project or facility specific, but the methodologies employed are potentially transferable to other applications.

CHAPTER 2. SYSTEMS OVERVIEWS

The overview is organized by life cycle function performed : type of costs considered, method of cost estimation, generation of financial indicators, and recomrnendations to vendors. This chapter also includes general recommendations to users of the software systems and tools.

COST-ESTIMATING SOFTWARE SYSTEMS

Cost estimation involves determining the quantity and predicting or forecasting, within a defined scope, of costs required to construct and equip a facility to manufacture goods or to furnish a service (Martin, 1992). Cost estimation is a standard exercise in business management, whereby rough and detailed estimates of project costs are estimated for various purposes. Cost estimation software systems may be used for capital budgeting and procurement purposes and for estimating and comparing different project costs. Cost-estimating software systems help users use, control, and manipulate data in a manner that simulates reality as closely as possible. They can do this much faster and more efficiently than any manual method. One of the most important facets of a cost- estimating system, from the point of view of incorporating environmental information, is its flexibility and versatility.

This chapter provides information on cost-estimation systems that are profiled in this book. Eighteen systems are included, of which nine were profiled and the rest summarized in abbreviated profiles. Most of the cost-estimating products profiled allow medium to high levels of user customization. However, the amount of flexibility required of a system depends on the application and the complexity of the project. The available software comes in a wide price range: from $995 for G2 Abacus to $5000 for COSTPRO. This wide range of prices reflects the varied abilities and features of these systems in terms of flexibility and user customization.

LIFE-CYCLE ANALYSIS (LCA)

Many industries define LCA as having an internal corporate focus and including costs incurred during capital investment, material acquisition, and cost allocation. Our definition goes beyond this internal focus to include other "life-cycle stages" that may not otherwise be considered. The stages start at extraction of raw materials and proceed to manufacturing and processing of materials, then to use and reuse of materials and products, and finally to waste management or recycling. Some products, such as Composer Gold, that have LCA components or modules use a definition of the term "LCA" that is not consistent with ours. These products are more consistent with "total cost assessments which is defined as "long-term comprehensive financial analysis of the full range of internal (private) costs and savings of an investment" (Spitzer et al., 1993). The LCA considered by a management or decision- support product vary with the products, processes, and projects being analyzed. Different project components will be in different stages of their individual life cycles. Most cost-estimating software systems are not designed for the specific purpose of life-cycle cost estimation and analysis. For simplicity and consistency it is indicated that these products are typically designed to consider the manufacturing stage of operations (construction, setting up process plants, etc.).

The nine cost-estimation software systems profiled in chapter 4 cover the manufacturing stage of the life cycle. Three out of nine additionally consider the use/reuse stage of the life cycle. Once again, corporations consider this stage with respect to their internal operations. The use/reuse stage will not, for instance, include costs associated with use/reuse of a product by consumers. Software that considers this stage is limited to use / reuse / maintenance costs that companies directly incur. All four stages were checkmarked in the case of one product-the *FAST* system. However, consideration of these stages is not integrated in one system. Different stand-alone

FAST models and systems address each of these stages. A complete range of environmental costs are not considered in any of these stages. Again, this system was not originally designed for LCA.

Many cost-estimation software systems can potentially be modified to include life-cycle cost information. For example, life- cycle stages can be distinguished through different Work Breakdown Structure (WBS) levels. Some products have restrictions on the number of WBS levels that can be considered, in which case this option may be limited. Users need to develop additional life-cycle cost information because this information cannot be derived from any of the existing cost databases. More importantly, information on environmental impact costs cannot be derived from existing cost databases.

COST ESTIMATING SOFTWARE

Cost-estimating software typically use databases of costs that are used to enter actual cost data or algorithms to estimate costs. Without these cost data or algorithms' an estimating system is essentially an application for managing and organizing data. It cannot generate reliable information without these costs. Most cost estimating products allow users to enter information on environmental costs in some manner. The important question is how these costs can be estimated, as described later in this section. Many of the cost-estimating software systems profiled have been developed for estimating construction and site preparation costs. The cost databases that vendors provide generally contain this type of cost information. Recently, databases such as the Army Corps of Engineers Unit Price Book and others have been attempting to broaden their focus by adding costs of maintenance and repair, waste management, and pollution control equipment. Thus, developers are beginning to incorporate some conventional costs and hidden costs associated with environmental management into their databases. However, the focus still remains on conventional costs, on items such as equipment, materials, labor, etc. All nine listed cost-estimation software systems help with estimating conventional costs. Three out of nine products additionally cover some aspects of hidden costs. The hidden costs typically include repair and maintenance costs and costs associated with waste management.

As mentioned above, databases typically do not provide information on costs of environmental compliance (e.g., permitting, monitoring, reporting), contingent environmental costs, and less-tangible costs. Users typically estimate contingency as a percentage of total estimated project costs, which does not necessarily factor in contingent costs that can arise due to liabilities associated with environmental damages or noncompliance with regulations. Image and relationship costs (a type of hidden cost) can include costs of environmental reporting and other costs incurred to boost corporate "green" images. These costs themselves are not intangible, but the benefits that accrue as a result of these often are. Databases and algorithms should be developed to provide users with estimates of hidden costs as well. Finally, there are external or social costs whose values would need to be imputed using various techniques (e.g., the social "cost" of air pollution as a result of siting a refinery). There are conflicting opinions on methods of treatment and estimation of less-tangible costs. Also, data on this subject may not be widely documented. Thus, incorporating external or social cost information into databases or project costs will require further research.

Estimating software typically comes with a specific format for entering and organizing cost information with or without the option for modification. These systems have not been designed to include all the cost categories we discuss in this handbook. Creating distinct cost categories for conventional, hidden, contingents and less- tangible costs may or may not be possible. Determining whether these costs can be made transparent (i.e., not clustered along with other cost elements) is important because cost elements that are not visible cannot be controlled. Changes in these cost elements through changes in project elements and directions should be identifiable. Many hidden costs have traditionally been lumped into overhead accounts, which made controlling them individually or identifying the cost drivers difficult. However, even if cost elements are transparent, users may not want to include them. For instance, businesses currently may not find it acceptable to make go/no-go decisions based on increases/decreases in total project costs arising because of including social (less-tangible) costs into the project totals.

It is useful to give users the option to compute and compare different project totals based on including or excluding different cost elements, including environmental costs in estimating software has two aspects: the manner in which these cost data should be *included,* and the development of unit cost data or algorithms to *estimate* these costs. Environmental costs could be included in cost estimating software systems in two broad ways.

Additional activities and resources: The activities or resources that create environmental costs can be included along with other elements that have related labor, material, and overhead costs. For example, permitting and waste disposal have costs associated with them and can be included as separate activities along with related conventional activities such as those under transaction systems or assets. Appropriate units of measure can be chosen and unit costs can be estimated for these items. However, a disadvantage in following this course of action is that the systems will not prompt users to include these elements, so they would include them only at their own discretion. Most estimating products would allow users to do this. This may or may not be a problem in some smaller systems that only allow a limited number of activities per project. If the system has sorting functions, users will be able to separately analyze these activities. Another way to keep these activities segregated is to include them in different WBS levels. Some systems do not have sorting functions, and some have only a limited number of WBS levels possible. Also, adding activities and WBS levels can affect the speed of operation in some packages (especially those that are Windows based).

Additional cost categories: Most systems have categories under direct costs (labor, equipment, materials) and indirect costs (overhead). Users may be able to create additional categories for hidden, contingent, and less-tangible costs. However, contingency is usually just computed for the project as a whole, not for some individual activities. Also, if users input numbers directly rather than take them from a specific database, the sources of these costs could remain ambiguous, thus hindering control. Some software systems, such as Hard Dollar EOS, allow a number of options for cost detailing and overhead detailing. However, most cost- estimating software allows only a limited number of categories (often not user defined), which users will typically use for conventional cost detailing. Both these options are related and the best strategy would probably be to adopt a combination of these two methods.

Cost Estimating Software

All the cost-estimating software profiled and summarized use either the unit cost or the parametric cost-estimation method. Some products' such as Composer Gold, Success, and Precision Estimating Plus, offer a combination of these two estimating techniques. The cost databases provided and the default algorithms in estimating systems also address conventional costs. As mentioned earlier, some additional conventional and hidden environmental costs are being added to these databases. However, most estimating software and databases provided by vendors do not give information on environmental costs or on methods for estimating them. Environmental costs included for examples regulatory costs, liability costs, and environmental impact costs. Some products give users the option to create their own database of costs or to add to the existing cost database. Existing databases (provided with or compatible with the system) may or may not allow users to append additional information. However, creating a database of costs for use in estimating environmental costs is not an easy task. The system's ability to use data from other software systems may be a consideration when creating a cost database. All the systems profiled are network compatible. Although most systems allow direct input of data, the source of the cost data is not always apparent. Here, annotation capabilities in the system can be helpful (e.g., project level, estimate level). Composer Gold (MCACES) allows users to attach information on the data source used for an estimate.

Another possibility for estimating environmental costs is to develop relationship models or algorithms within the system that would define the values of certain cost elements. For instance, users may be able to define a relationship saying that the permitting costs for a particular activity depend on the quantity of material X used. The values of these costs could depend on quantities or elements that are not included in the traditional analysis, in which case, users would also need to estimate these quantities. Again, this is not an easy task. Developers of parametric-estimating software systems such as FAST should research environmental cost estimation (e.g., estimates for hidden costs or contingent costs) using basic parameters.

Four primary financial indicators are identified in this guide: NPV, IRR, benefits cost ratio, and payback period. These indicators have advantages and disadvantages related to their simplicity and reliability. Conflicts in ranking projects can occur between mutually exclusive projects in the useful life, size of cash outflows, and timing of cash flows. The NPV method is generally recognized as being the most reliable method, because it measures the absolute value of all discounted cash inflows at the required discount rate. However, the

NPV is also one of the most difficult to calculate because it takes time and requires some financial analysis skills even if computer programs are available. The comparison of the IRRs of two or more mutually exclusive projects will not necessarily lead to the correct choice. The IRR method is biased in favor of projects that yield high rates of return on money "borrowed" early in the life of the project. Projects with large negative future cash flows are not attractive investments but yield abnormally high IRRs. Projects with future cash flows that switch from positive to negative (or vice versa) have multiple IRR solutions. A project with a higher cost can have a higher NPV than a cheaper project but have a much lower IRR (Winston, 1995). However, many industries employ the IRR as an indicator to meet the requirement that selected projects have returns beyond a predetermined "hurdle" rate.

The benefits cost ratio only looks at the ratio of total (discounted or undiscounted) benefits over total costs. It gives no indication of the increase in net wealth and is not a good method for ranking projects. For example, this indicator would give a project (assuming time = 1 year) with total benefits of $40 and total cost of $10 (BC ratio = 4) preference over a project that has total benefits of $2,000 and a total cost of $1,000 (BC ratio = 2) although the latter clearly increases net wealth more. This method is useful mainly as a rough indicator of whether project benefits exceed project costs. The payback period, whether discounted or undiscounted, ignores the cash flows occurring after the payback period. This method is often biased against environmental projects that typically have long gestation periods. The main attribute of this method is its simplicity and ease of calculation.

Many cost-estimating software systems (especially those that can be used to generate rough estimates) are used in capital budgeting processes. However, none of the systems profiled had any internal functions with which to perform financial analyses for comparing different projects or for evaluating different options in the same project. However, data export and import capabilities of the systems may allow transfer of cost data to other packages, such as MS Excel that perform financial functions.

Software vendors can take different steps toward meeting the challenges offered by the changing nature of business activity and requirements and toward addressing the need to make environmental costs part of the decision-making system. These steps include the following :

1. Ask for feedback on environmental costs that customers would like to see included.

2. Include environmental cost elements in the demonstration or sample projects provided to users in with packages or provided in training sessions.

3. Provide users with information on methods of estimating environmental costs.

4. Start including elements of environmental costs (at least private costs) in cost databases or algorithms to estimate environmental cost elements in parametric estimating software systems.

5. Include modules that allow users to compare alternative projects using standard financial analysis techniques such as NPV.

Developers of reports and methodologies (tools) on the subject of environmental costing can also collaborate with cost estimating software system developers to develop methods to include environmental costs in the estimation process. The comprehensiveness of costs covered will depend on objectives of the estimating process as well. Order-of-magnitude estimating, for examples will not require detailed inclusion of environmental costs. Similarly budget estimates and definitive estimates will require different levels of detail. Thus, appropriate methodologies need to be developed to estimate and include environmental costs in different stages of the cost-estimating process and also to support different cost-estimating methods.

P2 System Installation Scheduling and Cost Control : P2 project schedules need to be managed or monitored to prevent deviations from the project plan. Scheduling refers to assigning the estimated duration and desired start and finish times to each activity in a project within the overall time period required for completing the project (Martins 1992). Cost controls which forms part of monitoring and performance measurements involves minimizing costs and keeping expenditures within budget (track and calculate cost and schedule performance). Project monitoring and performance measurement are the systematic applications of methods and procedures to monitor performance against the plan (Parker, 1994). Of the eight "scheduling" software systems covered in this book seven have been fully profiled. Of the five "cost control" software systems included, four have been fully profiled. Scheduling software systems come in various levels of sophistication and include methods for inputting and editing project data, diagramming methods, and calendars. More sophisticated scheduling systems have resource scheduling algorithms that allow users to split activities or use alternate resources, for example (Winston, 1995). The level of customization permitted and the flexibility in specifying costs also vary. This difference is indicated by dividing the scheduling software systems into three categories: high-end, low-high end, and low-end scheduling. Users need to select these systems based on project size, complexity, and other specific requirements. The cost control software systems covered are typically designed to work with scheduling products and have features such as earned value analysis, trend, summary, and exception reporting. Two products are slightly different-Cost Time

Management and EasyABC Plus. The former uses a diagnostic technique called a "cost-time profile" to analyze and optimize project time and costs. The latter, EasyABC Plus, is a product for developing and modifying ABC models. It has applications beyond project tracking and reporting. It uses the ABC technique to analyze and compare costs at project planning and post-project stages.

LCA does not necessarily play a part in the scheduling and performance measurement functions. In these phases, the area of interest is limited to cost and time requirements during the course of a project. LCA can be employed for cost minimization and analysis during the project planning, analysis, and cost-estimation phases that precede scheduling. Thus, if a cost analysis application is used for project planning and analysis, it may be useful for it to have an LCA component. Life-cycle information can be factored in the same manner as described under the section on cost-estimating software systems.

Life-cycle cost information may not be relevant for scheduling and project cost control software systems, but considering a full range of environmental costs (that impinge on project costs) is definitely an important factor. For example, performance measurement may reveal that project costs for particular activities have exceeded the budget, possibly due to a hidden environmental cost such as environmental health and safety costs. If the cost driver cannot be identified, taking the appropriate corrective action and predicting or preventing future occurrences will be difficult. For this reason, activities related to environmental management need to be factored into project costs and schedules. The more sophisticated scheduling software systems usually afford greater flexibility, allowing users to account for environmental costs through a more significant breakdown of project activities. Sorting and ceding functions and WBS levels can be used to keep these activities and associated costs distinct. However, most of the systems (especially the lower-end scheduling softwares do not permit the inclusion of additional cost categories.

Scheduling and cost control and analysis systems are not designed for cost estimation. All cost information has to be entered by users or imported from cost-estimating software. None of the scheduling or cost control systems have internal functions that generate financial indicators, Most project cost control software systems, however, do perform returned or earned value analysis. Since the financial indicators listed are primarily used in the capital budgeting process, they are not relevant in the case of these software systems. It would be useful for products such as EasyABC Plus that are used in the pre-project cost analysis stage to have such functions.

RISK ASSESSMENT SYSTEMS

Determining risk is an integral part of project planning and estimation. Assessments of risks and uncertainties need to be factored into project cost estimates. Project risk management covers the measurement and interpretation of risk and includes techniques such as sensitivity analysis, decision trees, and Monte Carlo approaches.

Four software systems were fully profiled and one abbreviated. Of the products fully profiled, one is a generic risk assessment product (for company-wide application), and the remaining three have been developed specifically for estimating project risks and contingencies. The Woodward Clyde software (ERMIS) and tools for estimating environmental liability cask has been abbreviated because no specific information was available on the software used in the liability assessment process. This is because Woodward Clyde performs the risk analyses for their clients, with the software only playing a small part of the entire process. However, risk assessment software, similar to those described in the guide, plays a part along with expert information, in environmental liability costing.

Many generic risk estimation software systems, similar to MOUSE, are offered in the market and are usually very moderately priced.

Users can estimate uncertainties associated with static mathematical models developed for a variety of purposes (including environment-related). ICECAN and Primavera's Monte Carlo have more specific functions for analyzing project risks and contingency. For example, Range Estimating quantifies and ranks project risks.

Developers of REP software system recommend that users account for conventional, hidden, and contingent costs. These are not necessarily "environmental" costs. Indeed, none of the examples provided in the product documentation refer to environmental costs and the need to include them in risk analyses. Most project cost estimates include contingency estimates. ICECAN and REP are designed to produce contingency estimates based on variable data entered by users. However, contingency estimates do not usually account for potential environmental liability (contingent) costs, such as penalties and fines or legal expenses. These costs can play an important role in assessing the accuracy of project cost estimates. Manufacturers of products such as REP add that contingency estimates can factor in environmental variables. Thus, users will need to define these variables and include them in their analysis as desired. Only the Woodward Clyde software and tools are designed specifically for assessing risks or contingent costs as a result of environmental variables, They indicate that estimating liabilities may not be as simple as plugging in numbers and variable information into a risk/contingency estimating software system. Woodward Clyde points out that site specificity can make it difficult to design generic systems geared specifically for the estimation of environmental liabilities. Their method relics heavily on expert knowledge and site data, which they feel is more crucial to the accuracy of estimates than a software system alone.

Risk assessment software systems are not meant for cost estimation. However, they can be used to add to project cost estimates and refine them by incorporating the element of risk. They do so by evaluating the uncertainties in project cost estimates and by estimating cost contingencies based on variables identified by users. Moreover, there are no internal functions for generating financial indicators in any of these products. This function is beyond the scope of these software systems.

Developers of risk analysis software systems and methodologies for environmental risk management can take steps to provide users with information on how to factor in environmental considerations into their software systems:

1. Provide sample risk analyses and computations of contingent costs that include environmental variables.

2. Develop heuristics and algorithms and include appropriate probability distributions to account for these variables.

3. Developers of new methodologies and reports need to provide current information to users (possibly through software vendors) on methods of liability costing.

4. Developers of risk assessment tools or methodologies and organizations involved in liability costing research need to collaborate to demonstrate (e.g., through case studies) the use of available risk assessment tools and software for the purpose of estimating liability costs.

5. The accuracy of estimates determined through the use of available tools and software (with differing levels of expert knowledge applied) needs to be determined. This can help companies do trade-offs in deciding how much to invest in estimating contingent environmental

costs.

Developing expert systems (if possible) designed specifically for the estimation of environmental liabilities will take time primarily because research in this area is relatively new, and standard, accepted methodologies have not yet been developed for liability costing.

REMEDIATION SYSTEMS

Remediation products are developed with the specific purpose of handling various phases of environmental remediation projects. Software products designed for remediation projects typically handle estimating costs, tracking costs, or choosing appropriate remediation technologies. The products covered under remediation projects have been grouped under sub-categories for cost estimation and/or scheduling (three products), products for RI/FS (Remedial Investigation / Feasibility Study) planning and risk analysis (one product), and products for cost tracking (one product). Table 3-2 provides information on the software systems covered. The characteristics of these software systems are in most ways similar to more general purpose products that have been designed to address the same areas.

Life-Cycle Stages Considered (Remediation Project- Related Software)

Since all these products are designed to support various stages and requirements of remediation projects, they loosely cover the waste management stage of the life cycle. However, this information on this stage is not covered with the objective of doing an LCA or LCC, thus many of the project components could actually be in different stages of their individual life cycles. Other stages including use/reuse could also be partially covered.

Type of Costs Considered (Remediation Project- Related Software)

Most remediation products cover conventional and some upfront hidden costs, such as site preparation and other costs associated with other remediation project stages. RAAS is the only product that covers only direct conventional costs. These software systems need to be extended to include other cost categories as well.

The estimating software systems covered use the parametric estimation technique to estimate remediation costs or remediation technology costs. These software systems do not allow users to add their own algorithms to estimate other cost types. However, some do allow direct data input. This option, as mentioned earlier in the case of the cost-estimating software systems, can restrict cost control because it does not link with other project parameters. These products also restrict the addition of cost categories.

Chemical Tracking Systems

Environmental management and regulatory compliance software systems typically help companies keep track of chemical use, waste tracking, etc., for the purpose of complying with various environmental regulations, and more recently, for facilitating areas of corporate environmental management and environmental health and safety. The environmental management and regulatory compliance software systems are divided into two broad categories: facility management and environmental health and safety and waste management and tracking.

Six products are profiled under the first category, and 12 products under the second. Several systems offered by Environmental Software and Systems, Inc., include a number of repackaged software systems prepared for the EPA. Software systems such as

AWARE, DataPipe, and EMS have been designed for cross-industry applications, while others such as Gentrax and other Wixel products are designed for specific applications or for specific industry sectors and Federal facilities. These systems come in different levels of complexity and with different attributes, although the primary function remains the same. Many of the systems come with modular options that allow users to purchase additions only as required. Some products such as the Material Inventory Report System are completely limited to addressing regulatory reporting requirements and do not feature any additional attributes. Product enhancements and developments are based on user demand, changes in regulations, and anticipation of future changes. The only exceptions covered are real-time monitoring systems, such as the G2 Real Time Expert System. These systems have real-time and off-line modeling applications, which can be used for a variety of purposes including real-time emissions management.

In his article "Environmental Management Information Systems: New Tools for Measuring Performance," Chris FitzGerald (1994/1995) differentiates Environmental Management Information Systems from Environmental Information Management Systems. Indeed, most environmental computer applications and databases were originally developed to respond to specific regulatory requirements. He defines Environmental Information Management Systems as systems that were developed in efforts to manage the overwhelming volume of regulatory monitoring and reporting requirements. On the other hand, the Environmental Management Information Systems are organizational and technological systems developed to supply needed environmental information and data to customers engaged in improving corporate or institutional environmental quality (FitzGerald, 1994/1995). This is an important distinction, and this research indicates that software developers are increasingly including features in their systems to integrate environmental activities and programs along with other company activities. This is particularly apparent in the products for facility management and environmental health and safety. The smaller, less expensive products remain more limited in function.

FitzGerald (1994/1995) also points out the need to avoid data duplication in an organization, and the need to integrate environmental information with other standard company information. For example, hazardous material inventory data should be captured within the purchasing, distribution, and inventory functions. Similarly information on waste flows, manifests, etc., should be captured along with other data on company transaction systems. Although a majority of the systems covered have networking options, many are not designed to integrate, for instance, with company accounting systems. This can lead to separate management of environmental data. There are exceptions, however, such as AWARE, TINIA, and Plantware. However, once again, such features are more often seen in the larger, more expensive systems.

Most environmental compliance software systems consider only the waste management stage and the use/reuse stage of the life cycle. The manufacturing stage of the life-cycle and the waste management stage are not always included in the LCA systems. The terms have been loosely applied. Systems seem to cover the waste management stage when they primarily track waste products. The use/reuse stage is included only if raw materials are tracked, and if the actual manufactured products are tracked. Although a number of systems can track wastes back to specific manufacturing processes, they do not technically cover any information on the manufacturing stage of the life cycle due to the complexity introduced by chemical reactions in the processes that cause new chemicals to appear and feed chemicals to be consumed. Also, the information gathered in each of these stages is not intended to be used in an LCA. Thus, the information included is not strictly "life-cycle" data.

LCA can play an important part in such product/process design choices and methods for cost minimization, but it is probably beyond the scope of most of these products in their current configuration. However, the data collected through the use of these systems could be useful if such analyses were to be conducted. Many of these systems allow some information to be archived, although only in a specific format.

Some real time systems, such as the G2 Real Time Expert system and Neur-On-Line, have modeling capabilities that can be applied for cost estimation and LCC. However, these systems are sold as generic systems with no specific information provided on how users may develop such applications. Building in some basic modeling capabilities and algorithms may be useful in other products as well that allow users to estimate and analyze contingent costs and less-tangible costs. Relationship models can also be developed for relations between quantities of material inputs and waste produced (with costs factored in). Developing relationship models can be a useful extension of products that compute material balances and can provide useful support information for projects. These models can increase the cost control capabilities of the system by allowing users to evaluate the results of changes in the process, materials used, etc.

One major limitation associated with many of these products is that they do not cover a complete range of environmental costs and, therefore, cannot effectively address P2 and cost minimization. Many of the products track only some limited cost types, such as manifest costs, transportation costs, and others. Waste and waste cost minimization cannot be achieved if only partial costs, such as costs incurred for disposal of waste materials, are considered. Waste management costs can be most effectively minimized only when all costs, including processing, storage, health, and safety are considered. Ideally, conventional company costs and other hidden and contingent environmental costs need to be integrated to get an accurate perspective.

Apparently, some developers are beginning to realize the importance of full-cost tracking and are modifying their systems accordingly. These include software systems such as AWARE, EMS, TINIA, and Flow Gemini. Some products such as TINIA can integrate with company accounting systems, which prevents environmental costs from being managed in an isolated manner. Of the 17 systems covered in this section 15 have some cost fields (in their default formats). However, only two track product or production data, and three track capital and/or operations and management costs. One (Flow Gemini) plans to include these features in their next release. Some software products, such as Data Pipe, restrict users from customizing this area on their own. Some vendors add that this area would be developed only if their customers demanded it. Many will make appropriate format changes to add cost fields through specific customer requests. The drawback of this option is that some users will not make such specific requests for modification to vendors because they lack information in the area. Thus, this appears to be a transitional period with respect to treating environmental costs and activities as part and parcel of business as usual.

No products found/reviewed assist in including or estimating liability and less-tangible costs. Furthermore, most of the products do not currently have modules for project management. Some, such as TINIA, that do have this feature cover very basic aspects with the option for exporting data to other software designed more specifically for project management. Even these basic project management capabilities provide a useful feature. None of the systems reviewed facilitate using the cost data collected for future cost estimation purposes. This area will need more research because there is currently little consensus on methods of estimating environmental costs. AWARE tracks all conventional and hidden costs associated with environmental management in a company along with other features to facilitate P2 under current and anticipated EPA guidelines.

None of the environmental management and regulatory compliance software reviewed are used for cost estimation. Users supply all costs entered into the system. However, environmental management and information systems can track cost information that may be useful for future cost estimation purposes, particularly those related to environmental projects. Moreover, none of the systems reviewed include any financial functions primarily because they are not usually used for project planning and analysis. However, this will be a useful feature in products that have modules for developing and monitoring P2/waste minimization projects, such as AWARE and TINIA. Software developers would take many steps to make their systems more comprehensive. Short-term goals,include the following:

Recognize cost control as an important goal even for compliance-related activities.

Include a more comprehensive list of costs in their basic formats

Allow users to add activities or cost categories.

Include sample formats or examples in training sessions that comprehensively cover environmental costs.

Include modules that allow users to analyze, build, and manage P2 and waste management projects.These should include standard financial analysis techniques.

Provide features and mechanisms that will allow management of environmental activities to be integrated with other company activities. An example is the "request" method in HAZMIN,

where all material purchases are reviewed prior to approval.

Medium-long term goals could include the following:

Collaborate with developers of cost-estimating software methodologies for environmental costing, to determine whether and how cost information tracked using these products could be made useful for cost-estimating purposes.

Include cost-estimating modules that use historical environmental cost information and algorithms to estimate costs for future projects (both environmental and non- environmental projects).

Build in relationship models or modeling capabilities to enable estimation of less-tangible and liability (contingent environmental) costs and relationships between various input parameters and costs for environmental and waste management.

Developers of methodologies (tools) for environmental costing can work with software developers to formulate techniques for environmental cost estimation. Such research could fill the gap of information on historical costs related to environmental management. Many of these costs may be unique to particular companies based on their policies and methods of functioning. In such cases, companies may be able to use cost information gathered through using these products to build unique historical cost databases.

FINANCIAL ANALYSIS SYSTEMS

Financial analysis is an integral part of the capital budgeting and procurement process (wherein company funds are allocated to various projects), and relics upon various techniques and indicators to compare projects or present project outcomes in financial terms.

The products included under this section are either software-based spreadsheet programs that facilitate financial analysis using one or more methods or written presentations of methodologies. Six products, including one abbreviated profile, are included under this category, out of which three are software based applications. The level of detail in these software systems and tools varies widely. For example, the Pollution Prevention Benefits Manual presents a detailed methodology for data collection and analysis. The "Economic Analysis of Pollution Prevention," on the other hand, is a brief fact sheet with basic information on using the NPV as a method for analysis along with examples of cost elements that should be considered.

None of these systems profiled are meant for LCC. Indeed, at the financial analysis stage, only those costs that impinge on project costs or company costs are of interest. These products attempt to analyze costs by taking a total cost perspective. All these products cover the manufacturing, use / reuse / maintenance, and recycle / waste management stages of the life cycle. All the life-cycle data are not included. For instance, the use / reuse stage will not include data on costs associated with use of a product by consumers, and the waste management stage may not include environmental impact costs of a particular disposal option.

Some products, such as PRECOSIS and P2/Finance, are software based applications that include pre-defined formats where users can enter project cost information and perform financial analyses. None of the software systems allow users to develop models or algorithms to estimate less-tangible or liability costs (although P2/Finance allows users to input some liability costs). In all cases, the costs covered are restricted to internal company costs. However, users can include other social costs in the analysis in the case of manual methods and in P2/Finance (to an extent). The software-based applications often have limitations on the number of activities and cost categories that users can add. The important feature of software products and methodologies is that they alert users to include many environmental costs that are left out of traditional analyses.

Financial analysis products are not technically designed for cost estimation. Users have to supply all the cost data to be included, as these products have no default costs or cost databases. Since these are not estimating applications, they include no specific estimation methodology. However, the product may define the manner in which the cost information is to be entered into the system or presented for analysis. For example, PRECOSIS requires that users enter unit cost data for labor, materials, facilities, and waste management.

All five of the products profiled give users the option to calculate NPVs. Four out of five products give the option to calculate the IRR. None of the products profiled calculate this indicator. Three out of five products calculate the payback period.

Tools and software systems that use financial indicators need to recognize and acknowledge their relative advantages and disadvantages for the benefit of the users. For example, the NPV method is by far the most reliable method for assessing projects; it allows a fairer consideration of environmental and P2 projects. The Pollution Prevention Benefits Manual does not compare the pros and cons associated with using different financial indicators such as the NPV. The project time frame permitted by financial analysis software and tools is another important factor because many environmental projects have long gestation periods. P2/Finance allows user-defined time periods.

These software systems and tools can be more effective in the following ways:

Include methods of data collection and environmental cost

estimation in the manuals and software-based applications. This is a developing field.

Include algorithms (or the ability to include algorithms) to estimate environmental contingent and less-tangible costs.

Allow the computation of different project totals and financial indicators by including or excluding different cost elements.

Increase flexibility for adding additional activities and cost categories.

Include a comprehensive list of cost categories in the standard formats

Include discussions on the advantages and disadvantages of using one financial indicator versus others, in the context of environmental and regular projects.

Different versions of P2/Fi nance have been developed to cater to different industry sectors. Although generic applications are useful, products tailored to specific industries allow developers to include more specific cost categories and provide more guidance to users. These efforts should be encouraged.

ENVIRONMENTAL COST IMPACT ANALYSIS

LCC is the systematic process of evaluating the life-cycle costs of a product, product line, system, or facility by identifying life-cycle consequences and assigning measures of monetary values to those consequences (ICF, Inc., 1995). Impact analysis involves the assessment and/or quantification (not necessarily in financial terms) of environmental impacts that result from an activity or process. This section includes a variety of software systems and tools that serve the basic purpose of LCC and impact analyses. Some products, such as HAZMAT and Hazardous Material Life-Cycle Cost Model, have been designed for specific facilities and ends.

Fourteen tools and software were covered under this section, of which ten were fully profiled. The products range from simple spreadsheet applications such as the APACS LCC software demo to complex products such as EcoSys.

LCC models use costs through the life time of a project or material as a basis for analysis. As mentioned in Chapter 2 and in Section 3.1.1 on cost-estimating software systems, the business definition of the life cycle is usually different from the definition used in this handbook. Examples of tools employing this definition are the APACS LCC demo and the Life Cycle Costing Program. These LCC models typically include (other than up-front capital and operating costs) maintenance and repair costs and liquidation values. Of the ten LCC products profiled, only three consider all four life-cycle stages, and EcoSys attempts to create a complete life-cycle perspective. The Tellus Packaging Study includes all the stages except the use/reuse stage of the life cycle.

One of the reasons that most products do not perform a complete Lees consistent with this guide's definition is that the methodology is very complex and not as yet standardized. The Tellus Packaging Study uses our definition of the term but only does a partial analysis, not including all the life-cycle stages. EcoSys attempts to counter the problems associated with data collection by using an expert system and decision analysis to create a life-cycle "perspective," even though inventory data are included only from the manufacturing stage of the life cycle. TEAM provides users with the modeling tools to prepare and analyze a life-cycle cost model using all four life cycle stages. It includes a database on process information, but does not include any default formats for users to develop their models.

The LCC applications and case studies indicate the differences in interpretation and application of the term life cycle. None represent any standard or universally accepted methodologies for life-cycle analysis or LCC. However, they are building blocks that can be used to further develop approaches and applications in the area.

Methods such as the EPS Enviro-Accounting Method and the EcoSys Model (EcoSys) are examples of methods for environmental impact analysis. Although these do not actually quantify environmental costs, they present an option for including some form of qualitative analysis and including less-tangible costs into the decision-making process. HAZMAT is the only product that estimates contingent environmental costs based on pre-defined algorithms.

Developers of cost databases need to focus attention on the area of developing unit cost data (where appropriate) for environmental costs. Some efforts have included maintenance and repair costs and waste management and P2 equipment costs in databases. The reason other cost categories are not included is probably because the field is still in its infancy and has no consensus on methods for estimation.

Most of the products included do not generate any financial indicators. The exceptions are the APACS LCC product and the LCC program. TEAM does not have any of the indicators defined in this guide. However, it has all classic mathematical functions and user-defined formulas can be used to create specific indicators. Generating financial indicators would be a useful feature in other life-cycle assessment applications as well, because it would allow users to analyze the impacts of various alternatives on financial returns.

Factor in costs into applications that are designed primarily for LCA, allowing industry to use LCA for more intuitive decision-

making.

Include a full range of private and social environmental costs in LCC.

Once methodologies are more streamlined and accepted, life-cycle cost information could be incorporated into existing or new cost databases.

WASTE REDUCTION SYSTEMS

Waste reduction software and tools are developed for managing wastes and waste costs for specific facilities or as general guidance documents or software. This section includes some general purpose products such as the Facility Pollution Prevention Guide, EcoAccountingSM, and Industrial Waste Prevention. Although many of these products have been developed to serve specific projects or areas, the methodologies employed are often instructive for other efforts at developing similar methodologies.

Twenty products are covered under this section, of which 13 are fully profiled. Six of these products are designed for, or can provide, generic guidance for a range of applications. The rest can provide examples and guidance but are strictly designed for specific facilities and/or project types. This category also includes a subsection on proprietary software systems and tools. Only one product, the Rohm and Haas model, is included as a proprietary product. Additionally, a subsection covers two case studies; one for the application of life-cycle analysis, and the other, for activity based costing.

None of these products profiled were originally designed for or have modules for LCC. Their level of focus is restricted to internal company dynamics and costs related to pollution prevention or waste management. All the products fully profiled cover the waste management stage of the life cycle. Six products include data from the manufacturing stage of the life cycle.

Although the use/reuse stage is not totally ignored, none of these products include detailed information associated with this stage of the life cycle. Although methods to account for use/reuse is beyond the scope of most of the products covered, efforts should be made to integrate the use of these products with others that do perform LCC.

None of these products profiled cover all the cost types identifies in this guide. All these products cover some aspects of conventional and hidden costs (except one, that covers only conventional costs). However, the hidden costs, for instanced do not include regulatory costs and other voluntary environmental costs. Industrial waste prevention considers contingent costs but does not include a comprehensive methodology for estimating them. The only exception is the Rohm and Haas model, which uses the methodology described in the Pollution Prevention Benefits Manual and includes conventional, hidden (including image and relationship costs), and contingent costs. It does not include social (external) costs.

None of the products covered are designed for cost estimation, although the data collected by using them can be used for future cost estimation purposes. HCAS is essentially a cost database, which can further be used for estimation purposes. Users supply all cost data in all other products.

Most of the products covered, other than exceptions such as Industrial Waste Prevention and the Facility Pollution Prevention Guide, do not generate any financial indicators. Interestingly, the indicator of focus is the payback period, although others are sometimes mentioned, such as in Model P2 Opportunity Assessment, the payback period very often does not produce reliable results.

CHAPTER 3. COST ESTIMATING SOFTWARE SYSTEMS

This chapter provides a description of the cost estimating software systems reviewed by the authors. Wherever possible, cost information is given along with a means of contacting the supplier.

ComposerGold/MCACES

Prepared by: Building Systems Design (BSD)

Prepared for: Public purchase

Application: Environmental and Commercial Cost Estimating Software. The system has potential to be applied for estimating and including life-cycle environmental costs. The life-cycle module available includes the manufacturing and the use/reuse/maintenance stages of the life-cycle and considers only internal company costs. BSD-supplied databases currently cover environmental cost information on some upfront hidden costs such as site preparation, waste management equipment and technologies etc. Back-end, regulatory, liability costs, etc., are not covered.

Elements of control:

Cost control
Resource control
Estimating control
Schedule control
Scope control
Risk control

Development date and updates: First released in 1990. Updates every 1-2 years. The current release is version 6.0, released in November, 1994. The next release, expected in the first quarter of

1996, will be a Windows version.

Public availability: Available

Purpose and current use: BSD markets Composer Gold as an estimating system for large or small projects and can be used for most construction project work. Their client list includes architect and engineering firms, cost consultants, government agencies, and private corporations. The system has been approved by the Army Corps of engineers and the EPA for hazardous waste projects. The government version (MCACES) has some 1,000 users and Composer Gold has *500 users (mostly architects and cost consultants).

Cost information: The 1995 list price for a single user is $2,995. Additional modules and databases, such as the models capability and Life-Cycle Cost module are individually priced.

System requirements: The system runs only on DOS and can be used on a PC or a LAN. It requires a minimum of 512 K RAM and the software takes 8 Mb of space on the hard drive. The system is quite flexible in the type of LAN platform that can be used.

Software Summary

Size and complexity of projects: Can be used for any size project. The maximum number of projects and the activities per project are limited only by the size of the hard disk. However, the system has typically been used for larger projects.

Other compatible systems offered: Users can purchase additional modules for modeling capability, scheduling interface, material reprice, and modules for Life Cycle Costing and Historical Project Analysis. They can purchase required databases and digitizers through BSD as well.

Nonquantifiable information: Users can add notes while reviewing an estimate. Features of Composer Gold include the following:

· standardized WBS approved by Federal agencies
· project and site-specific labor and equipment rates
· modeling database to store historical data
· vendor price tracking

Users can enter project and site-specific labor and equipment rates. Besides being able to use built-in work packages and assemblies, users can define an unlimited number of assemblies, which can reduce entry time and effort. Up to six levels of project breakdown are available in the system. The modeling capability, WBS levels, ceding, and user-defined cost categories can be used to estimate and include life-cycle environmental costs in various ways.

Life-Cycle Stages Covered

Raw material acquisition

Manufacturing stage
Use/reuse/maintenance
Recycle/waste management

Composer Gold has a separate module for analyzing life-cycle costs. However, this uses a definition of the life-cycle (different from ours), where only private costs related to the project are included. This may or may not include recycling and waste management. However, in this case, the database provides information only on maintenance and repair (other than the manufacturing stage) of equipment. The net retention/disposal value can also be calculated. However, the costs of actually disposing of the material (on-site or off-site) may not be included in the database. The database is the MCACES maintenance and repair database (in the Life Cycle cost module), which allows users to account for costs associated with the use/reuse of building components (in the facility itself).

Including life-cycle information may be possible by defining different WBS levels for each stage and also by using the ceding capability. Up to 6 WBS levels can be defined and applying this option for different life-cycle stages could be limited by other requirements of the users. If users do not need to keep the life-cycle stages distinct, this information can be incorporated in existing formats. Users would need to develop cost data from a different source if full life-cycle costs need to be estimated.

Type of costs considered

Conventional
Potentially hidden
Contingent
External

The Unit Price Book contains some information on hazardous waste-related costs, such as for safety equipment and special construction for waste treatment. The system can be linked with various commercial databases that are used to estimate conventional costs and waste disposal costs. BSD is now offering an additional database for Environmental Restoration, the ECHOS database, which is brought out through a joint venture between Delta Technologies and Marshall & Swift. This database addresses environmental data from a technologies perspective and covers 4000 assemblies and 4000 unit cost items. In BSD, the models database contains a library of 44 facilities, which users can add to by copying other projects. The types of environmental costs covered include site preparation, construction, environmental technologies, equipment, etc. These databases do not contain cost information on many hidden environmental costs, and no information on contingent or external costs. Users may be able to incorporate these cost elements in different ways. Estimates will need to be made by users.

If users can develop the appropriate relationship models, parameters, and unit costs, the modeling capability of the system can be utilized to estimate quantities and costs for specific items that generate these environmental costs. Second, users can directly input costs into the estimate from their own sources. Third, users can define cost categories that address these items. Gold allows users to define and select up to 5 each of owner cost, direct cost, and indirect cost columns. If some environmental costs cannot be addressed using the predefined or conventional cost categories, users can add additional categories, such as 'external costs'. For example, an activity such as distillation, could result in an air pollution cost. This cost (which would have to be estimated by users) could be entered in a external costs column. Combinations of these approaches could be used to include information on environmental costs.

Method of Cost Estimation

The system uses *unit cost estimating and parametric modeling*. The company plans to additionally offer factor and range estimating at a later date. Users create the estimate in a project file by entering appropriate cost items to price each area of work. This can be done by tapping into the databases connected in the system or by direct data entry. The screen gives information about the source of the data (entered/changed by users or through a database). Descriptions, quantities, and unit costs can all be changed directly on the screen and users can control column names (e.g., labor, equipment) as required.

The system can import ready-made databases such as Richardson,

Means, Lee, Saylor, and others. It contains a hazardous waste database from the Corps of Engineers and EPA and its own database of unit costs. Users can also build a database, which can be used to include environmental cost information that may not fit in existing databases. The system can contain up to seven databases. The database types included are project (where the project estimate is built), models, assemblies, unit prices, crews, labor rates, and equipment rates. In the equipment rates database, data such as depreciation, cost of capital, and fuel and operating expenses can be entered. The labor rates database can contain information such as basic rates, overtime, travel expenses, etc. The labor and equipment can be grouped to form crews, which can be used for estimating as well.

The historical cost project analysis module has a database of over 1,800 projects that contains 3,100 facility types. The information is stored by facility name/size/cost per square foot/year built/location. The Historical Analysis Generator allows users to query the database to narrow project sampling to facilities of a similar type and size. A cost index can be used to compare the cost of the facility with other areas of the country or the world. The database can be appended with the users' own historical project files.

Users can use an MCACES maintenance and repair database (in the life cycle cost module), which provides information on ideal maintenance and repair schedules for different types of building components. The life cycle cost module also allows users to analyze energy-efficiency alternatives and differing fuel systems. It reports capital, operating, and energy costs, and net disposal/retention value.

The models module allows users to access and modify the formulas that calculate quantities. It can assist in planning complex features and tasks based on previous models. The system offers cost/escalation adjustment factors for the entire project by percentage. An escalation index can be used to compare similar items in different bids on one screen. A what-if recalculation feature can be used to compare different project scenarios. This can be a useful feature in comparing the effects that changes in different variables can have on environmental costs.

Generation of Financial Indicators

Net present value (NPV)
Payback period
Internal rate of return (IRR)
Benefits cost ratio
Other

Ability to Include Environmental Costs

User friendliness and flexibility

The scheduling interface computes overall project duration and end data, and users can define schedule relationships among activities using PDM or the ADM method. The program can export data to different scheduling packages and dBASE format files. Composer can interface with project management software such as Primavera, Open Plan, and Microsoft project. The system can use a digitizer to take off quantities. Data in the system are automatically compressed for storage and for copying to disks. On a network, users may exchange data and usage can be restricted, if desired.

The system is menu-driven and provides on-line help. The system's manual contains a section on estimating and quantity takeoff techniques as related to the program. The system can also detect logical errors and prevents users from deleting important data by mistake. Users have the ability to make customized reports. They can define an unlimited number of work codes, which can be used to sort and display data. Summary and detail reports can be prepared by sorting by chosen variables.

The product is the commercial version of MCACES, and unlike earlier versions it now comes with the latest version of the Army Corps of Engineers Unit Price Book (this was previously considered proprietary). BSD also offers the ECHOS database for Environmental Restoration, which addresses environmental data from a technologies perspective. Composer Gold is accompanied by a couple of sample projects and templates that users can use to build their own projects.

User-support

BSD provides user support through training and technical services and information dissemination, including new software reviews and newsletters. They offer a renewable support plan that covers hotline support and free updates of the software. The price of the plan is based upon the number of users-$495 for the first user and $195 for each additional user. Training classes are offered in Atlanta or on-site (depending on the number of users). A demonstration disk highlighting system features is available. However, it is not interactive and does not allow users to build-in limited sample projects.

Limitations

The system does not provide graphical capabilities such as time-cost curves, estimated/actual cash flow charts, and histograms. Users can open only one project at a time, although they may cut and paste from other projects. Each project can have only one WBS structure related to it (the program can support multiple projects). The system does not generate any financial indicators. The program does not have a menu function for exporting/importing data.

The system has not been designed for life-cycle costing (consistent with our definition) or for including all cost categories, such as regulatory, contingent environmental costs, and less-tangible costs. Information on some upfront hidden costs (e.g., use/reuse of building components, technology costs for ER) can be obtained through the MCACES and the ECHOS database. However, regulatory, back-end, and voluntary hidden costs, and contingent and less-tangible costs are not covered, thus leaving out less conventional areas in environmental cost estimating. Users would need to develop estimating methods and cost data to incorporate these types of environmental costs.

Basis for evaluation: Based on information and a demonstration disk received from BSD in March 1995. Additional information received through telephone communications with BSD representatives.

Contact information: Publicly available. Composer is the commercial version of MCACES.
Building Systems Design, Inc. 1175 Peachtree Street 100 Colony Square, Suite 1600 Atlanta, GA 30361; Peggy Woodal 404-876-4700; 800-875-0047; 404-876-0006 (FAX)

COSTPRO

Prepared by: INPRO,Inc. **Prepared for:** Public purchase

Application: Generic cost estimation tool. Has been used in environmental restoration projects, and has potential to be applied for estimating and including other life-cycle environmental costs. INPRO does not provide databases that would contain information on environmental costs.

Elements of control:

Cost control
Resource control
Estimating control
Schedule control
Scope control
Risk control

Development date and updates: Originally developed as the Los Alamos Laboratory Estimating System in 1973. Released as COSTPRO (PC version) in 1984. Different modules are updated at different time-period (not fixed). On an average, the user's module is updated 2-3 times a year, and the database manager, once a year. The latest release was on March 20, 1995. The company is planning a large revision of the product, which will include a number of enhancements, including a change to a Windows version. They may rename the software at this point. The release date is uncertain at this point.

Public availability: Available

Purpose and current use: The system was originally developed for the DOE. It has been sold to some 500 users, out of which approximately 200 are currently using the system. It is being used in all types of industry primarily in construction and environmental restoration and management projects.

Cost Information: List price: $5,300 (single user) · Database manager: $2,300.

System requirements: MS-DOS or PC-DOS operating system using an IBM-AT (or compatible), or a 386 with a Math co-processor. Can be used on a LAN.

Software Summary

Size and complexity of projects: Features include 12 completely distinct WBS structures, user-defined ceding, unlimited projects, and parametric estimation capability. The system can be used for small projects as well as for billion dollar construction projects. However, INPRO recommends that it be used for sophisticated data estimating. The system does not come with any templates, sample projects or default information, which may make it unsuitable for smaller users.

Other compatible systems offered: System is available in different modules.

Nonquantifiable information: Note-writing capability

COSTPRO is a generic cost estimating system that can be used to prepare cost estimates from the planning stage through detailed design. The program is written in FORTRAN. The code is a complete commercial rewrite of the Los Alamos National Laboratory mainframe cost estimating system. COSTPRO was developed by a nonprofit organization called Spectra Research Institute. It is maintained and marketed by INPRO. COSTPRO is actually a collection of costing and cost file utilities managed by a software driver. Estimators can use these functions to build an estimating methodology that suits their specific requirements. Features and attributes of COSTPRO include the following:

· when-adapted for preparing estimates for DOE projects (can be used to estimate all cost elements of a project, including assessment, design, and remediation)

· unlimited projects

· flexible series of report development menus

· extensive capability to apply markups for indirect costs, job conditions, productivity, and taxes

The system is presented to users to modify and customize in any manner they choose. This would give COSTPRO a lot of potential to be modified accordingly to include life-cycle environmental costs. The ability to define 12 completely distinct WBS structures, user-defined coding and parametric estimation capability, and the collection of costing and cost file utilities, would work to enhance its ability to effectively deal with these costs.

Life-cycle stages covered

Raw material acquisition
Manufacturing stage
Use/reuse/maintenance
Recycle/waste management

COSTPRO is not designed to consider more than one life-cycle stage, and this would typically be the manufacturing stage of the life-cycle. It may be possible to include life-cycle cost information by defining different WBS levels for each stage. Also, users have the option to define 12 completely different WBS structures to meet different requirements. This could allow the generation and comparison of different results. For instance, one WBS structure could include only conventional costs, another could include all private costs, and another could add social costs. COSTPRO's ability to support multiple projects could also be used. If the life-cycle stages do not have to be kept distinct, the information could be incorporated within existing formats. In all cases, users would need to develop environmental life-cycle cost estimating methods and cost information themselves.

Type of costs considered

Conventional
Potentially hidden
Contingent
External

Not unlike other cost estimating software, COSTPRO has been designed to consider conventional company costs alone. However, since users are provided flexibility in entering estimates, they can modify the system in a suitable manner to include other costs as well. Commercial databases (if any) can be used to directly input data, or users may create their own database of costs. However, until commercial databases include information on regulatory, back-end, and voluntary hidden costs, and contingent and external costs, users would need to develop this cost data themselves. Similarly appropriate estimation techniques would have to be chosen. Once the costs are available, they could be incorporated into the program in various ways.

If users can develop appropriate relationship models, parameters, and unit costs, the parametric estimation capability can be used to estimate quantities and costs of the specific items that generate these environmental costs. This can be helpful when exact quantities are not known. For example, a relationship may be determined between the size of a boiler (or some other cost driver) and the air pollution created, which generates a external cost whose value can be imputed for every unit of air pollution created. Secondly, users can directly input costs into the estimate from their own sources. The annotation capability can be used to attach an explanation of the source of the estimate. Third, user-defined cost categories can be created. They can be used to avoid entering some additional activities. For example, if users determine that using a particular piece of equipment requires a permit, or OSHA training (both hidden costs), they can enter these costs in a "hidden costs" column along with other conventional costs associated with purchasing and using the equipment.

Method of cost estimation

Three processing files are associated with developing a cost estimate within COSTPRO. The first is the database manager that provides the ability to draw data into COSTPRO from pre-established list(s) of items, activities, tasks, crews, assemblies, etc. The second is the take-off files that contain the unit rates, quantities, WBS, and other relevant information. Finally, the project files contain WBS and activity descriptions, codes, notes, cost factors, and other information associated with the estimate.

Users can also create conceptual estimates using parametrics, which can be further developed to detail actual costs. COSTPRO has a database manager that allows users to build custom databases of their own or use databases created by others. Thus, users can build custom databases that include various environmental costs estimated from different sources.

Generation of Financial Indicators

Net present value (NPV)
Payback period
Internal rate of return (IRR)
Benefits cost ratio
Other

Ability to Include Environmental Costs

User friendliness and flexibility

COSTPRO can combine inputs from commercially available databases (one or multiple) and can export data to scheduling programs. The system comes with a complete database manager for these applications. The accuracy of COSTPRO model estimates has been validated by the Los Alamos National Laboratory.

The system can simultaneously manage 12 distinct WBS structures for the same take-off without needing to re-enter the basic cost estimate take-off data. This allows users to use specific breakdowns for different requirements. The software manages all historical files. Users can customize the system in any way. They can create their own codes and charts, with which they can sort and resort the data in any combination. Users can generate exception reports by any combination of breakdowns. The system can manage large files without significantly slowing down operations.

User support

A detailed user manual provides instructions for model operations. INPRO can modify the model to meet specific customer requirements. Telephone hotline support and training classes for all the modules are available (on- and off- site). Updates are sometimes provided free of charge. Users who sign an update agreement receive all updates free of charge. A working demonstration disk is available for the user's module-this works on limited files and allows users to try out nearly all the features (not of commercial use).

Limitations

INPRO does not provide users with any databases (even commercial databases), although they do at times provide information on available databases. Although the system is extremely flexible, users have to build and adapt the system to suit their requirements. It does not come with any templates, or sample projects or default information, which can make it unsuitable for smaller users who do not require this level of detail or may want to be "walked through" an estimating methodology.

The system has not been designed for life-cycle costing or for including all cost categories. No environmental cost information or estimating methods can be obtained through INPRO. Thus, users would need to independently develop estimating methods and cost data to incorporate most environmental costs. The system is compatible with the MCACES database, which can provide users with cost information on some upfront hidden costs.

Basis for evaluation: Based on WEC (1992), and Youngblood (1992). Further information obtained through telephone communications with Dr. Alan Dudley (March 1995).

Contact information: Commercially available
Dr. Alan Dudley INPRO, Inc.
1613 University Blvd. N.E. Albuquerque, NM 87102
505-243-2800 505-292-9042 (24 hr/day number)

Freiman Analysis of System Techniques (FAST)

Prepared by: Freiman Parametric Systems (FPS), Inc.

Prepared for: U.S. Department of Energy, Office of Plans and Resource Management

Application: Parametric cost estimating tool for various functions. It estimates costs for items such as construction, equipment, software, and mining and is can be applied for conceptual to detailed estimating. FAST-E can be applied for predicting equipment costs for projects in consumer, commercial and industrial systems, commercial and military aerospace systems, industrial and chemical plants, and nuclear power stations.

Elements of control:

Cost control
Resource control
Estimating control
Schedule control
Scope control
Risk control

Development date and updates: Development began - 1980. The developer began researching and making parametric cost estimating systems in the late 1960s. The last release of FAST-E was in early 1995. The next major release will be the integrated system, and is expected in early, 1996.

Public availability: Available

Purpose and current use: Was initially developed as part of work done under a DOE project. Currently being used in various different industries.

Cost information: DOE has free access to the systems, but need to pay for time sharing services. The system prices can range from $1000 to $30,000. Detailed cost information can be obtained from FPS.

System requirements: IBM PC based. FAST-E, FAST-C, and FAST-CO can be run on IBM 286 or 386 computers but need a FAST hardware control locking device. All the systems require 10

MB RAM and 500 K (minimum) on the harddrive. Windows-95 compatible.

Software Summary

Size and complexity of projects: The system can be used for various project sizes and complexities.

Other compatible systems offered: Different systems designed for specific areas, such as estimating equipment costs, construction costs, etc, have been developed. FPS is working on providing the option for integrating these systems. The different FAST systems will be then available as modular options.

Nonquantifiable information: Information not available

Software Summary

The FAST system can estimate the costs of various design or schedule scenarios. The modeling system enables users to obtain early estimates, especially in the area of high technology. This is possible because the system is based on concept/design-related data, as well as the technological phenomena underlying design, which determine the manner in which costs can change. The Office of Assistant Secretary for Defense Programs in the U.S. DOE funded and used the following FAST items.

· **FAST-E:** estimates equipment costs, including costs for design, development, production, integration, installation, and operational and maintenance cost for any type of hardware or equipment.
· **FAST-C:** estimates construction costs, including operations such as site work, heavy construction projects, buildings, and utilities. It can calculate schedules, risks, and costs for all types of civil engineering

or construction programs.

· **FAST-CO:** estimates the cost of ownership of a project, or the total cost of a project from its conception, throughout its entire useful life. The model can estimate costs on a monthly basis for specific time periods up to 60 years. Coverage includes maintenance and support, enhancements, decontamination, utilities, operations, consumables, partial and complete decommissioning, etc. FAST-CO includes procedures for decontamination and cleanup of buildings, equipment, sites, or underground areas.

· **FAST-F:** estimates the funding profiles for engineering and/or production over a project's life and develops the cost and schedule penalty for changing a funding profile after a project is underway.

· **FAST-M:** estimates mining and subterranean project costs.

· **FAST-P:** measures the overall productivity and performance of a project. Can be used to determine the facilities and resource needs of any engineering and/or manufacturing operation.

Other programs include:

· **FAST'S:** a software cost model that uses the FAST (technology and performance) concept.

· **FAST-ABC:** the Activity-Based-Costing model used to properly allocate overhead costs to the projects that caused them. FAST-ABC allows users to define areas of indirect expenses. It organizes its algorithms based on how users define the character of the indirect expenses. It includes default settings for elements including setup and procurement costs, and transportation costs. There are no default settings for environmental costs.

All FAST systems, including FAST-ABC, use fuzzy logic to generate algorithms based on examples provided by users of the required structure using two or more sample projects.

Life-cycle stages covered

Raw materials acquisition
Manufacturing
Use/reuse/maintenance
Recycle/waste management

FAST has different systems that can consider these life-cycle stages. FAST-M, which can be used to determine mining costs, can consider the raw material acquisition stage. FAST-CO can consider the cost of ownership of a project, from conception through its useful life (for up to 60-year projects). This could potentially consider the remaining three life-cycle stages. It may also be possible to incorporate data from the first stage. FPS is currently developing an integrated system that includes all the FAST systems. This will allow users to consider all the life-cycle stages, using one system.

Type of costs considered

Conventional
Potentially hidden
Contingent
External

The system is mainly designed to include conventional costs. Hidden cost items considered will include maintenance and use/reuse costs. FAST-ABC can be used to make hidden environmental costs more visible. There are no default settings for estimating potentially hidden regulatory, voluntary, and other environmental costs. However, all FAST systems work using fuzzy logic. Thus, users can describe situations similar to the current project (for instance, by using old projects), and FAST will develop relationship models based on these examples. In this manner, all cost categories can be estimated by FAST systems. However, users will still need historical or sample projects that provide the required specifications and data needed to build these algorithms.

Method of Cost Estimation

FAST uses parametric cost estimating. Parametric modeling is a technique in which values are input as specific parameters. For example, for an estimate of the cost per square foot of constructing a building, square footage can be a parameter. Users can use such parameters to develop estimates from historical data. The database comes with the software but can also be defined by users. FAST also considers other factors such as design, location, and construction materials that can affect the cost. Furthermore, it considers the effects of factors such as earthquakes, tornadoes, or radiation on costs. The bottoms-up method of estimation can also be used. Site-specific construction information for these models was collected by Frank Freiman.

For example, projects in FAST-E can be defined using equipment weight, or users can define the product's technology and inputs that depict performance characteristics. ENTYPE is the factor that describes the mix of technological masses that make up the product. The program provides several procedures for calculating and parameters can be heat, motion, electrical, etc. Performance characteristics describe the equipment's functionality (inputs and outputs, performance requirements, such as gallons per minute). Projects in all FAST systems can use up to 5-level WBSs.

Generation of Financial Indicators

Net present value (NPV)
Payback period
Internal rate of return (IRR)
Benefits cost ratio

Others

Ability to Include Environmental Costs

User-friendliness and flexibility

FAST is user friendly and accepts individually tailored inputs. FAST-E has an extensive help library, and every input and output variable has a context-help capability (can also be used as a tutorial). FAST-E also has an extensive library of error messages. All FAST-E inputs and outputs, including accumulations and screen displays can be saved to output files named by users. These can be independently printed or edited using standard word processors. All the FAST models come equipped with self-contained tutorials.

FAST also estimates the costs of connective and support systems (piping, plumbing, wires). It provides risk measurements by displaying cost uncertainties for each cost segment. The system is especially useful in the early stages of a project, program, or product, when little design has been done. In addition to identifying principal features and relating them to cost, it quantifies intuition based on experience and can recognize the conditions that make situations different. It can produce detailed estimates as well. It includes tables and reference factors.

FAST-E, FAST-CO, and FAST-ABC may have the potential to be extended to analyze various areas in life-cycle cost analysis. A unique feature of PAST is its use of fuzzy logic. This provides a method for users to estimate environmental costs, by applying the required specifications. PPS is currently developing an integrated system that will include all the FAST models. This system will be PC-based, with the option to be used on a laptop and on networks. This will be available in around 1986. The system is expected to cost over $25,000. This system could have potential for being used for life-cycle cost analysis.

User-support

FPS offers a number of training packages, including on-site and off-site training. The cost of training is based on user-requirements. A working demonstration disk is available for FAST-E that allows users to analyze up to 3 studies.

Limitations

The FAST system models are relationship models; users need historical data to run these. That is, they need a basis and specifications from which to derive estimates; then they can proceed to modifying parameters to simulate different situations. FAST systems do not include any default information on regulatory, contingent, or less-tangible costs. If users want to extend the system to these other areas, they need to develop and apply the required historical information.

Basis for evaluation: The information provided here is based on WEC (1992), and Youngblood (1992). Subsequent reviews were done by incorporating updated product information and through telephone communications FPS in April, 1995.

Contact information: Available through Freiman Parametric Systems.
Freiman Parametric Systems, Inc. 2508 Iron Forge Road Herndon, VA 22071
Frank Freiman (President) Nancy Mauro (VP)
305-978-9597 800-892-4286 Internet: FASTE@AOL.COM

G2 Estimator

Prepared by: *G*2 Incorporated

Prepared for: Public purchase

Application: Flexible cost estimating system. G2 has been used for environmental projects that involve some form of engineering, construction or site preparation. It has potential for being developed to include environmental life-cycle information.

Elements of control:

Cost control
Resource control
Estimating control
Schedule control
Scope control
Risk control

Development date and updates: First copy was sold to Jacob's Environmental Co. in August 1988. On an average, there is a major release every 7 months, with several other enhancements in between. The current release of G2 is version 4.63, released in Spring, 1995. The next release will be the Windows version of G2 and is expected in the third quarter of 1995.

Public availability: Available

Purpose and current use: G2 Estimator is a flexible cost estimating system that is marketed for use in various disciplines including environment and remediation, construction, engineering, architecture, budgeting, and operations. About 25°/o of the users are doing environmental projects, 25% are for engineering, 25% for general construction, and the balance to various other companies. The total number of clients is over 1,000.

Cost information: The prices for stand-alone systems are G2 ABACUS-$995, Level 1-$3,000, and Level 11-$6,000. Prices are obtained from information package received in February 1994.

System requirements: The system can be used on a LAN. The minimum system requirements are a 286 microprocessor, 650K RAM, and MS-DOS version 3.3. The system can be used on a wide variety of networks including UNIX, Novell, and Banyan.

Software Summary

Size and complexity of projects: Small to large projects. However, primary clients tend to be in the construction business and Federal facilities with project budgets of 50 - 100 million. For instance, it has been used by the DOE for the Yucca Mountain project.

Other compatible systems offered: The system is available in three levels: G2 ABACUS, Level I, and Level II.

Nonquantifiable information: Users can enter notes at the project, worksheet, and line level.

The description given below mainly describes Level 11 (the highest level). However, all three levels contain similar features and differ in aspects such as the maximum number of worksheets and WBS levels possible. ABACUS allows only one WBS level. Level I allows five, and Level 11 allows 16. Abacus users are restricted to 50 worksheets with 99 items per worksheet, whereas Level 11 allows unlimited worksheets with a maximum of 999 line items per worksheet. G2 has several features for expediting and simplifying work and facilitating analysis. Features of G2 include the following:

· cost details entered in worksheets
· menus throughout the program that assist users in estimating
· flag displays to indicate need for any recalculation
· use of composite resources or assemblies, which can greatly reduce user inputs, because several cost items can be added to the worksheet at one time
· what-if analyses to view effects on project costs by performing functions such as entering changes in total values, quantities, and production rates or modifying them by a certain percentage

The WBS levels and the worksheets can be utilized for entering environmental cost information. At lower levels, the limits on the WBS levels and activities can restrict incorporation of these costs, because users will want to first use the existing space to enter conventional cost information.

Life-cycle stages covered

Raw material acquisition
Manufacturing stage
Use/reuse/maintenance
Recycle/waste management

The system is not designed to consider more than one life-cycle stage, which would typically be the manufacturing stage of the life-cycle. However users may be able to include life-cycle information in various ways.

Different WBS tree levels can be assigned to specific life-cycle stages. Up to 16 levels can be defined in Level II. The information can be entered by creating specific worksheets that detail activities in different stages. The ceding capability can be used to assign information from worksheets to be placed in different user-defined reports. Life-cycle cost information can also be detailed in one or more of the seven resource types.

Type of costs considered

Conventional
Potentially hidden
Contingent
External

G2 is designed to consider only usual company costs. Also, most of the cost information from the commercial databases provides information on usual costs. The database provided with G2 is organized in groups. With the relevant environmental cost information, users can create a distinct area in the database that details these costs. Users could include information on environmental costs using various methods.

The cost categories option allows users to specify types of costs used. The system contains five default cost categories that users can utilize or change. Seven types of resources would fit into these categories. Environmental cost information can be directly plugged into the estimate. Tag-type resources allow users to directly plug numbers into the estimate. Users can for instance, plug in ballpark estimates for various environmental costs for sensitivity analyses. The annotation capability (at the worksheet and line level) would help in specifying the source of the estimate. Second, if costs cannot be detailed through activities and conventional cost categories, additional categories can be created, such as potentially hidden or external costs. For example, an activity such as parts washing can have external costs due to solvent emissions or hidden costs due to OSHA compliance requirements. These could be entered as separate line items, or using the new categories, the hidden and external costs associated with parts washing can be entered at the estimate level. Users can also create separate worksheets that deal with project areas traditionally left out of analysis details. For example, they can create a separate worksheet called regulatory costs, where different costs arising from various environmental compliance requirements (e.g., manifesting, labeling) are detailed. Finally, environmental cost information can also be detailed in resources. A useful feature is that resource rates can be based on any units. Allowance-type resources allow application of markups or discounts to estimates. This capability can be used for sensitivity analyses using estimates of what environmental costs, such as contingent costs may add to total project costs (or other elements). G2 has nine separate cost items that can detail equipment resources, including costs associated with fuel, oil and grease, and operating costs. This is a useful feature in detailing costs of purchase and use for different equipment.

Method of cost estimation

G2 contains estimate worksheets, each of which represents a portion of the work to be performed. An estimate can contain an unlimited number of worksheets, each with up to 999 line items. A worksheet consists of three components. The first of these is a header that gives the definition of the worksheet as a whole. Detail line items contain information that includes resource descriptions, quantity, and unit price. Users can define the line items they want displayed. Details can be viewed at any point so that users can tell how the cost totals on the line items were calculated. Finally, there are the worksheet totals. Thus, users define various worksheets that describe their projects, and these contain product and cost information as required by users. There are separate areas for entering resource information.

G2 offers resource databases including the Army Corps of Engineers Unit Price Book, Richardson Engineering Services, NAVFAC and

R.S. Means. They also offer partial databases for different materials and processes including carpentry, electrical, and sitework. Unit cost information is retrieved from the database using resource groups and codes. Within G2 estimator, estimates may also be retrieved from other projects and databases. Users can also add new codes to the database and edit existing ones. The system facilitates using multiple estimating methods and the level of detail of the estimate can vary from conceptual to detailed. The methods that can be used for estimating include unit price, lump sum, unit man-hour, production, percentage allowance, crews, and assemblies.

Generation of Financial Indicators

Net present value (NPV)
Payback period
Internal rate of return (IRR)
Benefits cost ratio
Other

The system does not generate any financial indicators. However, it may be possible to run certain programs that do have financial functions from within G2.

Ability to Include Environmental Costs

User friendliness and flexibility

Other DOS programs, including Primavera, Lotus 123, and WordPerfect, can be run from within the G2 estimator. Information can also by exported to almost any software package that has import capabilities. G2 allows seamless integration with Primavera and Welcom Software Technology products. G2 allows for the input of take-off data electronically via a digitizer (digitizing hardware costs can be obtained from G2). Since the product is available in different levels, users can purchase a system that meets their requirements. G2 can utilize the Army Corps of Engineers Unit Price Book that includes some cost information related to hazardous and toxic, and radioactive waste.

Several features in G2 make it flexible. G2 can contain an unlimited number of projects and databases. Users can create customized databases by copying contents of groups in other databases. They can incorporate database updates simply by using a master option that updates all duplicate cost items. The latest version has a feature for spreadsheet editing whereby worksheets can be converted to a spreadsheet through a hotkey. Many features assist users with estimate automation, for example, by using production rates, duration, quantity or cost factors, and user- defined formulas. Users can also control several items in the system including data entry methods, displays, and monetary symbols. The inquiry option can give quick totals for the entire estimate, individual cost items, etc. G2 also allows standardization to suit internal company requirements, including user-defined report formats, database structures, and estimate templates. G2 allows the use of unlimited work code fields, which facilitates sorting of data and exception reporting. Users may also do line-item overwrites for work codes. G2 also features user-defined help topics and descriptions. One of the sample projects included in the demonstration disk is a landfill project.

User Support

G2 organizes product evaluation seminars and also provides support services for users. Maintenance is provided free of cost for the first year for the G2 estimator and for 30 days for the G2 ABACUS. This includes 24-hour (7 days a week) support, user hotline, on-line modem support, and all updates and enhancements. Users have an option to renew the maintenance agreement and the price depends upon the number of users. They also offer an unlimited site license program for $50,000 for an organization (and affiliates) for unlimited access for a period of 2 years (not less than 1.5).

Limitations

Although it may be used with a mouse, the system does not have the Windows structure. The new Windows package is scheduled for release on September 7,1995. G2 allows different levels of WBS structures depending on the level of the software purchased, but the WBS tree is not displayed. The demonstration disk provided contains demonstrations of cost estimations in projects in different fields including environment and remediation. It provides a good overview of the features and capabilities of G2. However, it cannot be manipulated by users to run their own sample projects. The system is offered in different levels, but not in different modules.

The system has not been designed for life-cycle costing or for including all cost categories. Although some resource types, such as equipment, allow users to detail some hidden costs such as fuel costs and operating costs, other potentially hidden environmental costs are not included. The system has not been designed to support less-conventional areas in environmental cost estimating. Users would need to independently develop estimating methodologies and cost data to include environmental costs into their projects.

Basis for evaluation: Based on information package and demonstration disk received from G2 in February 1995. Additional information was obtained through telephone communications with G2 representatives.

Contact information: Commercially available
G2 Incorporated Corporate Office 2399 5. Orchard, Suite 200 Boise, ID 83705

Craig Lindquist Phil Jenson 208-384-1300 208-384-0300 (FAX)

Hard Dollar EOS

Prepared by: Grantlun Corporation, "Construction Information Solutions"

Prepared for: Public purchase

Application: For managing bids and cost estimating. Designed for the construction industry and could be appropriate for environmental projects that have structures similar to construction projects. However, the focus of Grantlun is on the construction industry. The system has potential to include environmental cost information that would form part of the projects that the system supports.

Elements of control:

Cost control
Resource control
Estimating control
Schedule control
Scope control
Risk control

Development date and updates: EOS was introduced in 1987. The current release, EOS-95 (version 7.0) was released
in May, 1995. Updates are provided 2-3 times a year, on an average.

Public availability: Available

Purpose and current use: Hard Dollar EOS is designed only for the construction industry, for heavy/highway bidding, and building.

Hard Dollar EOS can be used for bidding roads, bridges, earthwork, and environmental work. Hard Dollar EOS/PE is designed for engineers and helps noncontractors produce fully detailed or historical-based estimates using tools tailored specifically to the engineering community.Hard Dollar EOS is used across the U.S. and Canada, as well as many foreign countries, by nearly a thousand estimators. About 1,500 companies are currently using the system. It can be used by Federal facilities as well. The Florida Department of Transportation is currently using the system. However, the system is currently targeted to industry.

Cost information: Single user prices: Level 1-$2,900 Level 2-$5,900 Level 3-$8,900
(Approximate prices based on conversation with Hard Dollar representative in March 1995).

System requirements: Hard Dollar EOS was primarily designed for the DOS environment but can be created for other platforms such as UNIX and Altos, as well. The system uses a fifth generation development language called ANGOSS SmartWare. It can function in either network (supported environments include Novell, LANtastic, Microsoft) or stand-alone environments. The minimum system requirements are:

IBM 486 (recommended)
40 MB hard drive
8MB RAM.

Software Summary

Size and complexity of projects: Hard Dollar EOS is available in three levels, according to users' requirements for projects. Different level systems can be mixed and matched on a single network. It features unlimited number of estimates and unlimited estimate size and detail, and unlimited resource files and unlimited resources per file. They have been used for projects with budgets of below $1 million and those over $75 million.

Other compatible systems offered: Hard Dollar EOS manages the estimating and bidding side and can be linked with another Grantlun product, PXS, which handles the building (project management and control).

Nonquantifiable information: It is possible to add unlimited notes at the bid-item level.

During the bid-closing, EOS has the ability to automatically determine the best combination of quotes, using predefined award logic. Detailed budgets can be prepared, sorted by accounting or for the field. EOS automatically updates estimates after changes at any level. Features for estimating and resource management include the following:

unlimited number of estimates and unlimited estimate size and detail

5-levelWeS

features for combining the work of multiple estimators into one project

unlimited resource files and unlimited resources per file. Seven resource types and rates can be plugged in or detailed with 23 different rate elements.

variable costs and rates for different resources and items such as labor, fuel costs and sales tax

over 50 standard reports with custom reports available

Costs can be detailed in seven methods. Users can plug in bid item, activity, and detail-level costs: list resource requirements: use crews and set production; break work down into tasks; and/or proportionally reproduce previous work. EOS can use six project overhead detailing methods-general expense, abnormal equipment expense, project management expenses, overhead percentage add on, labor fringe add-on and taxes and fees in dollars or percentage of the job. All this gives users some flexibility in entering estimates.

Higher levels add additional features. For example, level 2 adds a historical cost library and facilitates automated import of historical costs from other sources. Level 3 allows users to import equipment rates from the Dataquest Equipment Cost Estimator (DECE) and allows them to query and index historical costs. The capability of EOS to support unlimited estimates and resources can be a useful feature in allowing users to enter information on environmental costs. Since DECE is meant to support construction projects, it probably will not include information on equipment for pollution control, waste management, etc. However, even construction projects will doubtless have environmental components including various regulatory requirements, contingent, and less-tangible costs. This type of information should begin to be included in databases providing information on construction costs.

Life-cycle stages covered

Raw material acquisition
Manufacturing stage
Use/reuse/maintenance
Recycle/waste management

The system has not been designed to consider more than one life-cycle stage, and this would typically be the manufacturing stage of the life-cycle. However, users may be able to include life-cycle information by using various functions.

Life-cycle cost information could be included by defining different WBS levels for each stage. The project size could limit using this option because a maximum of 5 levels can be defined. However, an unlimited number of activities can be entered. If users do not need to keep the life-cycle stages distinct, the information can be incorporated in existing formats, using cost detail worksheets. The unlimited resources can also be a useful feature. For example, users may be able to detail costs of one resource in various life-cycle stages. If an activity uses that resource, users could define that it uses four different resources (that is, the same resource in four different stages) instead of one. This would allow users to enter all life-cycle costs of using a resource, while keeping the costs incurred in different stages distinct (at the resource level). Needless to say, these costs will be difficult to determine, and academia has not yet reached a consensus on estimating methodologies. All such cost data would need to be developed by users.

Type of costs considered

Conventional
Potentially hidden
Contingent
External

The system is designed to consider only conventional costs, in various levels of detail. Environmental cost data could be included within EOS in various ways.

Users can directly plug in environmental costs into the estimate from their own sources. Distinct activities generating environmental costs can be entered, and additional cost categories can also be created. EOS allows users to add and modify fields and add an unlimited number of cost categories in addition to the ones in the system. Thus, users can add additional cost categories, such as hidden and external costs. For example, use of a particular construction material could require a tax, protective equipment, or testing. It may be possible to define some of these costs within existing formats (e.g., equipment sales tax can be detailed), enter them as additional activities (e.g., stormwater management), or enter the information in an additional cost category. That is, at the estimate level, along with entering conventional costs associated with a particular activity, a hidden cost category could detail additional costs due to requirements for testing or protective clothing. This would work toward keeping such costs visible, rather than hidden in overhead accounts. Environmental information could also be entered along with resources. The system can also support unlimited resource files (and resources per file). Use of some resources could result in certain environmental costs, and information on additional resources (traditionally not included) could be detailed. One could conceivably add natural resources that would be "consumed" as a result of various activities into the list. For example, if a construction project involved dredging and building over a wetland, the wetland can be assumed as a resource that is imputed a certain unit cost (e.g., value per unit area), which is used up as a result of the project.

Method of cost estimation

EOS has five different direct cost estimating methods that include direct input, list of resources, crew, man-hour (tasks), and spreadsheet (user defined). Thus costs can be detailed using users own resource rates or by looking into historical costs. EOS has six different project overhead estimating forms, including an unlimited spreadsheet that is defined by users.

Hard Dollar also offers some options to assist in cost estimating. The Dataquest Equipment Cost Estimator (DECE) is an equipment charge rate estimating tool that contains Dataquest's industry standards (or the users' own). The Estimate Detail Database (EDD) is an electronic encyclopedia of detailed bid items developed by professional estimators. Each item includes production rates and resource requirements and noted which can be used in Hard Dollar.

Generation of financial indicators

Net present value (NPV)
Payback period
Internal rate of return (IRR)
Benefits cost ratio
Other

No financial indicators can be generated. However, it may be possible to export EOS estimation data to other software systems that have functions for financial analysis. EOS has a built-in spreadsheet that can be linked with other spreadsheets that do have financial functions.

Ability to Include Environmental Costs

User friendliness and flexibility

EOS and PXS can share data and integrate the bidding and building process. EOS has been verified to be compatible with many accounting systems such as J.D. Edwards and DENES. Data from other software programs, such as digitizer systems, various spreadsheet systems, and dBASE, can be imported and used in EOS. EOS also has a built-in scheduling system (added in Level 4), data from which can be exported directly into Primavera format. Data created in EOS can also be exported for use in many other software systems, such as Lotus 1-2-3, accounting systems, and Word. EOS also has a built-in telecommunications system. The development language of Hard Dollar, ANGOSS SmartWare, makes all these features possible, because it provides full integration between built-in spreadsheet, database, word processing, and communications software subsystems.

EOS allows flexibility in estimation (e.g., an extensive breakdown of labor and equipment rates) and features a built-in Lotus-compatible spreadsheet system and a built-in database system. The system includes templates that can be used for creating custom reports. The next release of EOS (Level 3) is expected to include programming tools that will give users the option to reprogram the system to include as many optional fields and categories.

User-support

Hard Dollar customers can take advantage of training sessions and 24-hour support (as part of the annual renewal fee) offered by the company. Hard Dollar systems have been designed to allow nonexperts to use them with ease. The price includes free support and maintenance for one year, which includes toll-free support and all updates and enhancements. After the first year, users have the option to renew the maintenance contract. The price of this contract varies. Users can avail of a limited demo is offered for a specific time period.

Limitations

EOS features have been designed with the specific aim of targeting the construction industry. This can limit its ability to support all unique project types and data, such as life-cycle cost information. The literature does not discuss the accounting for "full costs" (repair, maintenance, etc.) within a company, but it may be possible to use the system with databases that provide this type of information. The system does not have library functions for ceding various variables whereby exception reports sorted on one specific variable can be generated.

The system has not been designed for life-cycle costing and for including all cost categories. The databases currently provided by Grantlun do not include information on environmental costs. Users would need to independently develop) estimating methods and cost data for including environmental cost information into projects.

Basis for evaluation: Information package received in November 1994, and telephone communication with Grantlun Corp. representative in March 1995.

Contact information: Commercially available
Grantlun Corporation "Construction Information Solutions" 2055 E. Centennial Circle, Suite 107 Tempo, AZ 85284
Ryan Brady 800-637-7496 602-897-6939 (FAX)

Precision Estimating Plus

Prepared by: Timberline Software Corporation

Prepared for: Public purchase

Application: Cost estimation. Designed for the construction industry and could be appropriate for environmental projects that have structures similar to construction projects. However, the focus of Timberline is on the construction industry. The system has potential to include environmental cost information that would be part of the projects that the system supports.

Elements of control:

Cost control
Resource control
Estimating control
Schedule control
Scope control
Risk control

Development date and updates: Developed - 1986. 1-2 upgrades per year. Timberline has been in business since 1971. The current release is version 5.9, released in Fall, 1994.

Public availability: Available

Purpose and current use: Primarily targeted to, and used in, the construction market. Timberline has about 6500 clients, each with 2 to 3 stations, on average.

Cost information: Base system: $1400. Average cost of add-on modules is $1,200 to 1,500.

System requirements: The system is DOS-based. Can be used on a network. Supported networks include DOS-compatible LAN, IBM-PC networks, IBM-PC Token ring, IBM LAN Server, Novell Advanced Netware, and MS LAN Manager.

Software Summary

Size and complexity of projects: The system is sold in three different levels so that it can be suitable for small subcontractors as well as for large projects. Can handle a certain amount of project complexity for particular project designs.

Other compatible systems offered: System is modular in design. Timberline also offers a variety of databases to users. The older system is called simply 'Precision Estimating' and does not have many of the features of Plus. Another version. Precision Extended, has the maximum features.

Nonquantifiable information: A pre-determined length of notes can be written in at every phase level.

Features of Precision Estimating Plus and Extended include the following:

· individual costs can be changed by percentages/flat amounts
· work packages (assemblies) for speedier item takeoffs
· display of total costs, total units, and unit costs immediately after takeoff
· automated pricing of crews-preset or user-built crews can be used
· all dimensional takeoff information stored-enables users to check back on how quantities were derived

· what-if analysis and model estimating capability in Precision Extended
· up to 12 of 20-character user-defined WBS codes can be attached to each item in Precision Extended

A custom report writer is available among additional software. Default waste percentages can be assigned to each item.

The system is designed for use in the construction industry, and the databases available through Timberline provide conventional cost information in this area. However, even construction projects have environmental components including various regulatory requirements, contingent, and less-tangible costs. This type of information should begin to be included in databases providing information on construction costs.

Life-cycle stages covered

Raw material acquisition
Manufacturing stage
Use/reuse/maintenance
Recycle/waste management

The system is not designed to consider more than one life-cycle stage, which would typically be the manufacturing stage of the life cycle. Users may be able to incorporate information from other life-cycle stages, but this information may have to be grouped with other information due to the limitations in defining WBS structures (up to 3 levels). However, WBS codes in Precision Extended could be used to generate reports detailing information on distinct life- cycle stages. The package can support unlimited projects and activities. This could be a way to incorporate life-cycle information. Separate projects dealing with distinct life-cycle stages could also be created.

Type of costs considered

Conventional
Potentially hidden
Contingent
External

The databases provided by the company do not provide users with information on anything other than usual costs. However, users can build their own database, or modify those provided through Timberline. Thus, they may be able to add information on certain types of environmental costs in the database.

Users can **directly enter cost data** into the estimate and attach explanations in notes. Users may be able to incorporate environmental cost information derived from other sources, but they **cannot define any new cost categories.** Precision provides information only on five conventional cost categories including labor, material, equipment, subcontractor, and other. Conventional environmental and some upfront hidden costs could be included among these categories or detailed among the **99 user-defined subcategories.** For example, a subcategory could be created for waste management equipment. The "other" category could be used to detail a particular type of environmental cost, such as potentially hidden costs. However, it is more likely that users would have other, more conventional uses for this. The system can support an unlimited number of estimates. Therefore, a better way would be to incorporate environmental costs estimated by users through more thorough **activity (or item) breakdowns.** Although reporting is a hidden cost, it can be entered as an item that has labor and other costs (fees) associated with it. The modeling capability in Precision Extended could be applied for estimating some environmental costs for which relationship models can be developed (using conventional parameters). For example, cost of waste management can be linked to plant size. **User-defined markups** could also be used to model the effects of project cost increases as a result of environmental costs, such as contingent costs. Cost data and estimating methods would need to be developed by users.

Method of Cost Estimation

Estimates can be built and summarized on a spreadsheet screen. The system uses **unit cost estimating, and Precision Extended adds some parametric estimation capabilities.** Precision has logic, trigonometric, and yes/no formulas. All Precision core products come with a working database, and users have the option of obtaining additional databases for various construction specialties through Timberline. With any of these, unit cost items, crews, productivity, and other data are automatically pulled into Precision's on-screen spreadsheet as users do takeoffs.

Generation of Financial Indicators

Net present value (NPV)
Payback period
Internal rate of return (IRR)
Benefits cost ratio
Other

The system has a series of internal mathematical functions but has no financial functions. However, it may be possible to export data to a spreadsheet that has financial functions.

Ability to include environmental costs

User friendliness and flexibility

The system is menu driven and uses an interface developed by Timberline itself. They are planning on introducing a Windows version in the future. The system is compatible with other databases, such as R.S. Means. A software called Timberline Bridge, along with other options such as the Microsoft Project and Primavera integrators, provide the capability for interfacing with Lotus, WordPerfect, MS Project. Primavera scheduling software, AutoCAD, and others. Timberline also offers a Precision digitizer.

The program can support unlimited projects and unlimited activities within projects. There is also no limit on the number of items, divisions, subdivisions, subcontractors. etc. The program is menu-driven and features on-screen help messages.

User-support

Timberline offers users a free "Total Care" plan for a year, which includes product maintenance, all updates, and toll- free user support. They also offer training (-$1,400 - 1,500) in various locations in the country. Other user support includes regional meetings, conferences, newsletters, and consulting services. Demonstration disks (slide-show) of program features are available. No working or interactive demonstration disks are provided.

Limitations

The system has a pre-defined fixed number of cost categories (e.g., labor, equipment), and users cannot define any others. The WBS tree can have a maximum of three levels. The system has no mouse support. Although users can include information on other costs as well (unlimited activities within projects), they cannot define these as separate cost categories.

The system has not been designed for life-cycle costing or for including all cost categories. The databases currently provided by Timberline do not include information on environmental costs. They do not give cost information on waste management, repair/maintenance, etc., either. Users would need to independently develop estimating methods and cost data for including environmental cost information in projects.

Basis for evaluation: Based on telephone communication with Timberline representative in March 1995. Additional information obtained through information package and demonstration disk.

Contact information: Commercially available
Timberline Software Corporation 9600 SW Nimbus Beaverton, OR 97008
John Geffel 503-626-6775 800-628-6583

Success

Prepared by: SoftCOST, Inc.

Prepared for: Public purchase

Application: Cost/parametric estimating software system. Success has some potential to be
modified to include life-cycle environmental costs.

Elements of control:

Cost control

Resource control
Estimating control
Schedule control
Scope control
Risk control

Development date and updates: First release: 1990. Release 2.2: March, 1995 Windows '95 version: expected Fall, 1995 Updates are released every 6 months, on an average. 'Decision trees' will be available as a value-added product in -August of 1995.

Public availability: Available

Purpose and current use: There are - 900 users of the system. It is being used in Federal facilities (including the DOE), power industry, pharmaceutical, and mid-large sized architectural and engineering firms.

Cost information: Single user: 51,495. Multi-user discounts available.

System requirements: The system requirements are: 80386 Windows compatible PC / 4+MB RAM
/VGA resolution monitor, and a mouse / MS DOS version 3.3 (or higher) / MS Windows 3.1 (or higher)

The system can also be run on a network, in which case any MS Windows compatible network may be used.

Software Summary

Size and complexity of projects: Large and small projects. However, displaying project complexity can be limited.

Unlimited WBS can also be used for Activity Based Costing (ABC).

Other compatible systems offered: Success Toolkit for Windows.

Nonquantifiable information: The system has a feature for adding notes to any element in the estimates.

Success is a Windows-based estimating and cost management system. The system allows users to prepare conceptual as well as detailed estimates. Features of Success include the following:

· graphical interface
· drag and drop capability to move and copy WBS components and cost items
· unlimited number of WBS structures, which can be viewed in the form of a tree
· pre-defined parametric models that can be used or modified through visual tools
· sets of variables for cost-drivers can be defined with default values, upper and lower limits, and complete descriptions
· dynamic or manual system totaling
· links drawings and images linked to project levels or detail items
· WYS/WYG report writer

Success can be used directly with GTCO roll-up digitizers. Success can be shipped with RS Means, Richardson, and/or the US Army Corps of Engineers UPB (free of charge) databases. It also supports all TRACES RACER models (free of charge).

Life-cycle stages covered

Raw material acquisition
Manufacturing stage

Use/reuse/maintenance
Recycle/waste management

The system is not designed to consider more than one life-cycle stage, and this would typically be the manufacturing stage of the life cycle. The system has been designed for estimating construction costs, and any cost information received from SoftCost would probably be limited to this area. Success also supports all TRACES RACER models that can be used for estimating remediation-related costs.

It may be possible to include life-cycle information by defining different WBS levels for each stage. An unlimited number of WBS levels can be defined. If users do not need to keep the life-cycle stages distinct, this information can be incorporated in existing formats. Users would need to independently develop cost data and estimating methods from different sources if full life-cycle costs need to be estimated.

Type of costs considered

Conventional
Potentially hidden
Contingent
External

The system is designed to consider conventional costs, and also can include some aspects of hidden costs. US Cost has developed 'decision trees' in SUCCESS for environmental cleanup activities. Success can use the TRACES environmental assemblies to estimate hidden costs associated with environmental technologies. It is possible to include other environmental costs within Success.However, cost data supporting estimation of all environmental costs is not available.

If users can develop the appropriate relationship models, parameters, and unit costs, the modeling capability of the system can be utilized to estimate quantities and costs for specific items that generate these environmental costs. For example, users may be able to develop relationship models linking one or more parameter (preferably a conventional item for which data is available) such as quantity of paint, to generate quantities of an item, such as air pollution (using a chosen unit, with an imputed value) that result in an environmental cost. Second, users can directly input costs into the estimate from their own sources. The detail item fields in Success are fixed, and the program computes totals for a pre-defined set of categories (e.g., direct, indirect). However, users could use three fields called "other" to detail some environmental costs. For instance, users could say that use of a particular type of paint results in conventional costs and other costs, such as hidden or external costs, each with a calculated, imputed or estimated value. Equipment costs can be detailed to include values of fuel costs, repair and maintenance costs, operating costs, etc, by using the markup feature for each designated equipment in the equipment table.

Method of Cost Estimation

The system utilizes unit cost and parametric estimating techniques. Success is compatible with databases such as R.S. Means, Richardson Engineering Services. Army Corps of Engineers UPB, and Navy CES. The parametric estimating feature allows users to prepare estimates using broad project parameters to select assemblies and calculate quantities. Users can use predefined models in Success, modify them, or create their own using the visual modeling tools. The parametric estimator automatically calculates quantities for line items or WBS levels. *The unlimited WBS in Success is used to build alternative cleanup processes (from the UPB and other databases) into decision trees. Users can also build models on the fly, that is, users who have no ceding skills can also define relationships among different items and use that to link different WBS items.*

Users can simultaneously access multiple projects or databases. Unlimited project resources include labor, equipment, crews, crew adjustments, materials, contractors, and currency exchange.

Generation of Financial Indicators

Net present value (NPV)
payback period
Internal rate of return (IRR)
Benefits cost ratio
Other

The Success Toolkit allows users to add the Visual Basic application which allows seamless interface with other products. This can be used to transfer data to products that have financial functions. The report writer also allows users to establish some equations.

Ability to Include Environmental Costs

User friendliness and flexibility

The system has an on-line help feature. The R&R report writer for Windows comes as an integral part of Success. The Windows and menu-driven structure is a plus point. New features of version 2.2 include Success Scheduler that automatically generates a schedule in MS Project using OLE 2.0, Success Toolkit Version 2.2 (optional) that allows users to seamlessly connect Success to CAD, CAE, scheduling, project management, and corporate management systems. Success Model Estimating allows budgets to be prepared in minutes with as little as one data selection.

SoftCost offers the U.S. Cost Inc. database and the Army Corps of

Engineers UPB at no cost. They also provide the TRACES environmental assemblies database at no cost. US Cost has developed 'decision trees' in SUCCESS for environmental cleanup activities and is in the process of more fully automating the decision process using the Success toolkit for Windows.

User-support

Success provides users with working demonstration disks that permit limited application of program features to a sample project. Annual maintenance options and telephone support are offered. Training is available for the basic system, parametric estimating, building/parametric models, and advanced users (including reporting).

Limitations

The WBS tree allows for users' convenience, but project details cannot be viewed at every level. Users can view totals at each WBS level, such as labor, materials, and overheads, but cannot tell the source of these estimates or view the detailed itemized cost breakdown. The detailed breakdown can be viewed on the spreadsheet associated with the final breakdown at each WBS level. The system has not been designed for life-cycle costing and for including all cost categories. Users would need to independently develop estimating methods and cost data for including environmental cost information in projects.

Basis for evaluation: Information package and demonstration disk received in February 1995.

Contact information: Commercially available
SoftCost, Inc. 5605 Glenridge Drive Suite 850, Atlanta, GA 30342
Steve Williams 800-955-1385 404-252-1385 404-252-3748 (FAX)

TRACES

Prepared by: U.S. Army Corps of Engineers, in conjunction with various software developers.

Prepared for: U.S. Army, Navy, and Air Force

Application: Umbrella linking includes automated cost engineering systems, and their databases. Scheduling software can also be linked to TRACES.

Elements of control:

Cost control
Resource control
Estimating control
Schedule control
Scope control
Risk control

Development date and updates: The system was referred to as TRACES (that covers the whole system) only in March 1993. Some portions of TRACES (MCACES) are required to be used in federal facilities.

Public availability: Available to the tri-services.

Purpose and current use: In 1990 the U.S. Army, Navy, and Air Force agreed to pool their resources and develop a single automated cost engineering system. TRACES cannot as yet be fully integrated, as there are still some stand-alone modules. However, the objective is to completely integrate the system.

Cost information: Private users can obtain TRACES by subscribing to the National Institute of Building Sciences, Construction Criteria Base (CCB) at a price of $750 for a one year subscription (includes 7 CDs). Government users and government contractors can obtain licenses through the US Army Corps of Engineers.

System requirements: The software requirements will vary depending upon the modules used. The systems are IBM-PC based. The CCB is single-user and PC-based. Some of the systems (such as MCACES) are network compatible.

Software Summary

Size and complexity of projects: The system is directed to large projects. However, it can be used to manage projects of various sizes and complexities, constrained by the individual flexibilities of the constituent modules.

Other compatible systems offered: The system is modular in design. All the modules are not integrated, primarily due to their diverse functions.

Nonquantifiable information: This will vary with module. For example, MCACES (described in the resource guide as Composer Gold) has an annotation capability.

The TRACES system intends to provide a user-friendly cost engineering platform in a standard environment, that will provide the cost engineer the tools to prepare, review, and maintain all types of cost estimates. Software for scheduling construction can also be linked to TRACES. The system includes the following modules:

- **Detailed quantity takeoff (QTO) module (MCACES):** For

detailed construction cost estimates for military, civil works, and HTRW programs. Includes a project database and supporting databases.

· **parametric estimating module:** For the preparation of cost estimates based on past designs, and less than complete information. Estimates can be transferred to MCACES.

· **Dredge estimating system (CEDEP):** Stand alone program for the preparation of pipeline, hopper, and mechanical dredge estimates. Includes two support data files.

· **Programming cost estimation module (PC COST):** Personal computer version of the DD 1391 Processor cost estimate generator. Preparation of programming, or budgetary costs based on historical costs, or from existing detailed or parametric cost estimates. Interface across MCACES databases, HAG databases, and TRACES parametric system databases.

· **Life-cycle cost module:** A stand-alone program for the analysis of system design alternatives. Includes a maintenance and repair (M&R) database tailored for Army buildings. Import from MCACES estimates. Conducts LCC in accordance with statutes and regulations, and calculates present worth individual building or facility components.

· **Historical cost analysis generator (HAG):** A stand-alone program to collect, store, and analyze historical cost data for facilities, sitework, and for HTRW remediation projects. Uses a standard WBS to track costs by type, location, size, and time.

· **Scheduling interface module:** A stand-alone program for scheduling of MCACES cost estimates. Data export capabilities.

· **Risk analysis systems:** A stand-alone program to help identify high elements of risk in a project, measure their impact on the cost of the project, and manage risks through contingency assignments. HAZRISK is one of these systems.

· **001391 Processor:** An interactive program that assists users in the preparation of programming construction cost estimates shown on the DD From 1391, Military Construction Project Data (used by DOD agencies).

TRACES includes several databases and files. These include the Unit Price Book (UPB), a models database, an assemblies database. Work Breakdown Structure (data file providing separate hierarchical WB master structure), Cost escalation index. Area cost factor index. Facility cost ($/SF), and the Historical Cost Analysis System (HCAS). The QTO database consists of approximately 22,000 items covering a full range of data supporting military construction, large civil works estimates, tunneling, and highly toxic waste and radiation waste cleanup estimates. This database is also used with the maintenance and repair database to provide a life-cycle cost analysis.

Life-cycle stages covered

Raw materials acquisition
Manufacturing
Use/reuse/maintenance
Recycle/waste management

The life-cycle cost module is designed to include all relevant facility costs incurred over the life of the facility. The life- cycle cost module consists primarily of a comprehensive and detailed maintenance and repair database, developed from the U.S. Department of Defense's engineered performance standards down to the maintenance and repair subtask level, and software that automatically computes the life-cycle costs of design alternatives on the basis of established DOD criteria and standards.

Type of costs considered

Conventional
Potentially hidden
Contingent

External

The method considers only conventional costs and some aspects of hidden costs. For example, the HTRW databases, and the HCAS databases will include some information on hidden costs. Also, the MCACES maintenance and repair database has information on hidden costs.

Method of Cost Estimation

Empirical estimating relies on historical data for cost development. It is based on the identification of building type, using historical unit cost (per square foot) and adjusting it by several standard factors including location, cost growth to the midpoint of construction, size, building sophistication, and a local site adjustment for each project that accounts for factors such as local contractor competition, site overcrowding, availability of crafts, etc. The quantity takeoff module is the U.S. Army Corps of Engineers' microcomputer system (MCACES). It is a step-by-step cost estimating tool that considers contractor layering, overhead, profit, and all indirect costs.

Generation of Financial Indicators

Net present value (NPV)
Payback period
Internal rate of return (IRR)
Benefits cost ratio
Others

Ability to Include Environmental Costs

User-friendliness and flexibility

The objective is to develop an integrated system that is user-friendly and works in a standard environment. The flexibility will vary, depending upon the characteristics of the constituent modules.

User-support

The TRACES group (Huntsville Engineering Division-Automated System Branch) provides user-support for the software programs through operations, maintenance, and hotline telephone support. Also, the Cost Engineers Bulletin Board (CEBBS) allows cost engineers to exchange information between districts, divisions, and authorized commercial vendors.

Limitations

The QTO database may provide data applicable to DOE projects. The system is not currently available for use by private users. It is primarily focused on construction and engineering projects. Currently, no data on contingent environmental costs or less-tangible costs is estimated. A full range of hidden costs is not considered.

Basis for evaluation: The information provided here is based on a profile of the tool presented Weitz et al. (1994), and information obtained from the US Army Corps of Engineers in June, 1995.

Contact information: HQ, US Army Corps of Engineers
Attn. CEMP-EC 20 Massachusetts Ave, NW Washington DC 20314-1000
202-761-1240 202-761-0999 (FAX)
CD-ROM subscription service National Institute of Building Sciences
202-289-7800

ABBREVIATED PROFILES OF COST ESTIMATING

ASAPMS Housekeeper

ASA Project Management System (ASAPMS) Housekeeper is a cost-estimating system that is targeted to architects, engineers, and contractors. The program combines estimating and scheduling through internal cost control and scheduling modules. It interfaces to other databases such as RS Means. The estimating program transfers takeoffs to a spreadsheet. It allows tracking of manpower requirements, project costs, individual activity costs, profit estimating reports, and cost variance reports (differences between actual cost and budget).

The program comes in three levels:

Level	Maximum activities	Cost (single user-system)
Educational	200	$400
Regular	800	$1200
Professional	8000	$4000

Multiple projects can be supported, with a maximum of 299. Additional cost categories can be added through specific requests to ASA. The vendors can modify the system to meet specific customer requirements.

Another system, called the ASAPMS Project Scheduling Library is also offered. It includes 7 independent products: two books and 5 computer programs. The five computer programs are ASAPMS LOTUSI23, BARCHART and MILESTONE METHOD, CRITICAL

PATH METHOD, PRECEDENCE METHOD. and PROJECT EVALUATION AND REVIEW TECHNIQUE (PERT). BARCHART. CPM, and PDM are made up of three modules: TIME ANALYSIS, RESOURCE CONTROL, and COST CONTROL, including earned value analysis. All these programs are IBM-PC based, and work on client server LAN, or single user environments. Single user prices range from $149 to $4000. These programs also come in three levels.

ASAPMS was first released in 1969, with system updates released every January. The current release is Version 26, January, 1995.

Minimum Hardware: Developers can tailor the ASAPMS system to almost any hardware configuration. IBM or VAX main frame with hard disk; will not run on a Macintosh. Can also run on micro systems (IBM PC, XT, AT, and PS/2 or compatibles). The educational and the regular levels require DOS with 1MB of core memory. The Professional level uses a Pentium, with 4MB of core memory. The system can also be run on a network.

Contact information: P.0. Box 3397, Church Street Station NewYork, NY 10008 (212)321-2408 (Phone and FAX) Andrew Sipos

Basis for information: Excerpts from Watson (1995), and telephone communications with supplier.

Cost Engineer

This modest cost-estimating system can be learned within an hour. It comes with a nationwide database that can be accessed through "keywords." The program is targeted to contractors, architects, engineers, and to environmental estimators for asbestos abatement and dig and haul site work. Suppliers provide their own database, which is updated yearly. It is ZIP code oriented with 31,000 items. The database includes construction costs, as well as costs for asbestos removal/abatement. Environmental costs such as costs of well installation and soil tests are included. The database can be updated and customized.

The program downloads into Primavera. Data can be exported to any program that can read Dbase formats, including Chase, Lotus, and Oracle.

The program comes with a complete (actual) demonstration project that people can use to build their own projects. Additional cost categories can be added by users. This includes repair and maintenance costs. An unlimited number of activities and projects can be defined. WBSs of up to 4 levels can be defined per project.

The program was first introduced in 1985, with updates released once a year. The current (1995) release is version 7.1. DOE has tested this program in West Virginia, and Federal agencies have used it for all projects in New York City. In addition, the company offers an international database with quantities in metric units. Users can generate estimates using foreign currencies since it factors in exchange rates. The supplier offers phone and consulting support, and on- and off-site training.

The approximate cost of the program is $495, plus $495 for the database. The metric database also costs $495.

<u>Minimum Hardware:</u> IBM PC, using DOS with hard disk or equivalent: will not run on a Macintosh. The program can also be run on a network. The Windows version is expected in 1996.
Cost Engineering Technologies P.0. Box 5474 Route 46E
Parsippany, NJ 07054 (201) 335-1707 201-263-1059 (FAX) Ken

Brown

Basis for information: Excerpts from Watson (1995), and telephone communications with supplier.

Cost Estimating

This is a spreadsheet template that provides a methodology for estimating costs of pollution control equipment. Cost methodology includes factors method, capacity change method, year of installation method, and both capacity and year change. Data for air pollution sources is included. Data on other types of pollution control can also be entered. The spreadsheet can be used as part of the BAT, BACT, or 'top down' economic analysis of new air sources.

Types of costs considered include labor, purchase price, and utilities (electricity, steam, and water). There are some cost figures in the spreadsheet that can be modified by users. No database of equipment costs is provided.

The cost of the program is $39. The system is available in MS DOS and Macintosh formats.
Environmental Software and Systems, Inc. PO Box 1182 ,Bowling Green, OH 43402-1182
(419) 353-8540 (419)-353-8540(FAX)

Basis for information: Information received from supplier.

Enhanced Cost Estimating Relationship Program

The program permits users to develop cost-estimating relationships (CERs) by inputting cost and descriptive data for known systems. The CERs are used to develop equations to predict costs of similar systems. Users enter information on multiple projects that are similar. The program uses regression analysis to develop algorithms linking costs with key variables. A maximum of 15 variables can be defined.

The program can predict the 95% confidence level of the cost estimate of a new system. Escalation factors can be stored. This permits an estimator to make an estimate based upon historical cost data. The database is user supplied.

The program was developed - 1989. There have been no changes or enhancements for the past few years. This cost estimating system is targeted to estimators, architects, and engineers. The program is available free of charge to DOE and DOE contractors. However, the program is not currently available through NTIS or OSTI.

Minimum Hardware: IBM PC with hard disk or equivalent: will not run on a Macintosh.

IT Corporation (formerly PEI Associates, Inc.) 4425 West Airport Freeway Suite 350 lrving.TX 75062 (214) 570-2000 Bill Gallagher

Basis for information: Excerpts from Watson (1995), and telephone communications with supplier.

Esceomate Piping/DWV and Sheetmetal/Duetwork Estimating

Esccomate is an estimating package for sheet metal and piping subcontractors. Features include single key-stroke item entry, voice input takeoff, digitizer control of computer, roller scale input, different material specifications, automatic calculations of hangers, excavation, couplings, bolts, gaskets, welds, and duct Joints. Space is providcd for two labor units and two priccs for cach itcm. Matcrial quantity takeoffs can be used to print purchase orders. Two databases with 6,000 to 7,000 items (Harrison Publishing House database for piping) are used. The approximate cost for the sheet metal module is $1,650 and for pipe module, $3,020.

Only upfront installation costs are estimated. Other costs categories, such as maintenance and repair costs, are not estimated, and cannot be included in the analysis. The program was released in 1967. It has not been updated since 1985.

Minimum Hardware: IBM PC with hard disk or equivalent. It will run on a Macintosh with an MS DOS board.

Environmental Systems Construction Company P.O. Box 461 FortCollins, CO 80522 (303) 484-8200 Dick Scoff

Basis for information: Excerpts from Watson (1995), and telephone communications with supplier.

PACES (Parametric Cost Engineering System)

PACES is a parametric cost-estimating system for conventional construction projects and is targeted to government and conventional construction estimators. Predefined and documented engineering relationships link the primary parametrics to detailed engineering quantities. Quantities and most model assumptions can be changed by users. PACES estimates are recognized by the Congressional Budget Office as accurate up to the *35°/o* design stage, and they have been validated to be within *71/2**o of final, actual costs. Estimates can be refined at later stages of design. It is comprised of 68 parametric building models that are grouped into 11 building categories. In addition, PACES has supporting facilities models, renovation models, and a general conditions, overhead and profit model. The renovation models were a major enhancement in the program in its release as PACES. The database used is the Army Corps of Engineers Unit Price Book. Estimates can be exported to MCACES gold.

The original system was released in 1984, on a mainframe. The PC version 1.01 was released as TRACES (the parametric Building Module) in 1991. The first complete stand-alone version was released in June, 1992. PACES Version 3.0 (the first time the system is referred to as PACES) was released in June, 1995.

The software can be obtained by attending a training class in Niceville (Florida), that covers the training and materials, system software, and documentation. The cost of training per individual is $750, with a class size of a minimum of 5 individuals and a maximum of 12. PACES is available only to federal facilities and government contractors.

Telephone hotline support is offered to users. No demonstration disks are available.

Minimum Hardware: IBM PC 286 or compatible with 15 MB or hard drive. It can run on a network. However, it is not designed to be network-compatible, and thus will not allow concurrent access.

Delta Research Corporation 1501 Merchants Way Niceville, FL 32578 . Manager, Facilities/Computers/Training: Robert Smith Steve Essig (904) 897-5380 (904) 897-5388 (FAX)

Basis for information: Telephone communications and information received from supplier in June, 1995.

Pulsar

The program is targeted to the government and commercial

construction industry. This is a cost-estimating software package that provides unit cost estimating. Reports can be issued to include various kinds of information about an estimate and can be exported to various spreadsheets, project management systems, databases, and graphics packages. Custom labor trades and equipment codes and rates can be created, and custom crews and assemblies can be built. Full type equivalent costing is available through the use of wage tables. Software can accommodate up to six Work Breakdown Structures (WBS) per line item, each having up to 4 levels. The WBS is user-defined and is in CSI format.

The project can support multiple projects and activities, that are limited only by the size of the hard disk. Costs can be changed to reflect local conditions. A user cost datafile enables users to include their own cost information. Also, cost subcategories can be created. However, additional cost categories cannot be created. The database is supplied by RS Means or can be user-defined. Database options are available to suit different project sizes. The company also plans to offer the Marshall and Swift ECHOS database as an option.

The software was initially developed by RS Means, who are now in partnership with Software Shop Systems. The system was released in 1989, with one general update each year. The Windows version 1.0 was released in June, 1995. The latest DOS version is 3.03, and was released in January, 1995.

Technical and maintenance support, including toll-free user support (8.30-8.00 Eastern time) for one year is included in the price of the program. Documentation and tutorials are also included. Demonstration disks are available. The DOS version has an interactive demo, limited he database and the line items (a maximum of 10). The Windows version does not as yet have an interactive demo. 2-3 sample estimates are included with the software. However, these serve only as tutorials and users cannot build their projects from these. On-line help is available.

The program is under General Services Administration (GSA) contract so DOE gets a discount on its purchase: 1st copy-$2,625 ;2nd to 10th copy-$1,575 ; 11 or more copies-$1,312

The price for private users is $3500. There is no quantity discount. The GSA price of the network version of the program is 3150 for 2 users, with $500 for each additional user. The retail price for private users is $4200, with $500 for each additional user.

Minimum Hardware: IBM PC (286), DOS 3.0, and 535K of free memory. Windows version available. Will not run on a Macintosh. Will support almost any network, including Novell, Windows for Workgroups, Banyan, etc.

Software Shop Systems, Inc. 1340 Campus Parkway Wall. NJ 07719-1973 (800554-9865.Ext. 3023 908-938-5631 (FAX) Sean O'Sullivan

Basis for information: Excerpts from Watson (1995), and telephone communications with supplier in July, 1995.

Questimate 8.1

The program is targeted to estimators of process plant projects, revamps, or retrofits. The modeling system calculates process equipment costs and costs for complete process plants. It permits the evaluation of alternatives by inputting the design specification. The equipment design parameters are input and the package produces costs, bill of materials, and labor requirements, which can be used for scheduling. The design can be conceptual or detailed. This is an assembly-driven package. The database is built into the system. The system can estimate costs of refits, retrofits and revamps as well. The supplier is developing more models of plant/equipment options that will allow Parametric input to create detailed equipment specifications and estimates.

The program was first released -1986. There is one cost update and one technology update each year. The latest release is version 10.0 (released in 1995).

The package is designed strictly for cost engineering and does not include project management capabilities. Users cannot define project work breakdown structures and activities. No sensitivity analysis can be performed within the system itself. No additional cost categories, including regulatory, liability, and less tangible costs, can be included in the estimate. Project Management capabilities are included in ICARUS' latest product: ICARUS Project Manager.

The company offers the system through a licensing plan where the cost of the single user or network version is $6900 for the first year and $2900 per seat (person) per year thereafter. The license has to be renewed annually. The maintenance fee is included in this. ICARUS offers consulting support, training, documentation, and an on-line tutorial. Demonstration disks are available, but these are primarily marketing tools, and are not interactive demonstrations of the program itself.

Minimum Hardware: 286 PC with 20 MB. LAN Manager, Novell Netware and DEC Pathworks network versions of the software is available.
ICARUS Corporation One Central Plaza 11300 Rockville Pike Rockville, MD 20852 (3011881-9350 301-881-2542 (FAX) Rick Geritz (Director, Sales)

Basis for information: Excerpts from Watson (1995), and telephone communications with supplier.

Richardson RACE (Rapid Access Cost Estimating) System

RACE is a computerized unit cost estimating system that can use up to five databases. It is targeted to general construction and process plant estimators. It has a spreadsheet-type worksheet, which is used to build estimates in small details (I ,000 items per detail and 1,000 details per project). Individual ceil access is provided, allowing for easy entering and editing of data. It can interface with Primavera and allows file importing and exporting.

It has a "Search and Scan" capability, allowing easy location of data within the databases. RACE uses Richardson's databases with over 100,000 line items. The database can be adapted to estimate maintenance and repair costs as well. However, it is primarily meant for estimating costs of new installations. No environmental costs are included in the database.

The predefined cost categories (material, labor, hours, subcontractor, equipment..) cannot be changed. There is an additional category, termed as 'other' that can be used to define other cost types. Also, costs can be placed in a summary area (outside the estimate).

No graphical WBS can be created. However, unlimited user-defined codes can be created to link detail worksheets. The system does not have Parametric estimating capabilities.

The program was first released in -1991. Updates are released approximately once a year. The current release is version 3.0 (1994) and version 4.0 will be released in August, 1995. RACE has users in the construction and manufacturing industry.

The approximate cost of the single-user program is $895, with the database available from $295 to $4100 for the main database. The cost of the network version is $995 (base cost), with the main database costing $4550. Multi- pack/site licenses are available. On and off site training is available. The cost of on-site training is $795 for two days for each group (with additional costs for travel and stay). Off-site training is available for $795 per person. Free (not toll-free) telephone support is offered. The program comes with documentation and includes on-line help.

Minimum Hardware: IBM PC or compatible. An 8088 processor, 640K RAM, one floppy drive, and a hard disk drive with 5-18 MB free disk space. Can run on networks. Supports Ethernet, which includes Novell, Lantastic, Banyan, and others.
Richardson Engineering Services Inc. 1742 South Fraser Drive P.0. Box 9103 Mesa.AZ 85214-9103 (602) 497-2062 (602) 497-5529 (FAX) Jack Helfrich

Basis for information: Excerpts from Watson (1995), and telephone communications with supplier in June, 1995.

StreetSmarts

This unit cost estimating system is targeted to heavy construction contractors (highways, bridges, dams, demolition, water and sewer systems). It is not meant for generating conceptual estimates. Estimators can set up rate and crew libraries, historical cost data, and an account structure such as C51 to prepare the estimate. The estimate can be used to track cash flow, prepare bids, and perform risk analyses. Reports can be issued in 15 versions, including total cost, cost for each work operation, expenditure analysis, revenue analysis, and resource demand. The database is user-supplied and fits into the PICK database management system in any UNIX environment. The system supports the analysis of an unlimited number of projects.

Additional modules can be purchased from the company to perform features such as scheduling and cost control. No databases are supplied by the company. Users can create their own databases, or link the system to available commercial databases. There is no Parametric estimation capability.

The program was developed in 1993, with updates offered 2-3 times a year. The current release is version 3.52, released in April, 1995.

The minimum cost of the program is $20,000, including hardware and training. The cost ranges anywhere from $20,000 to $250,000. The supplier offers training seminars. Some consulting services and on- and off-site training are offered. No demonstration disks are available.

Minimum Hardware: IBM compatible PC on a LAN with a UNIX server. UNIX emulator software installed on PC. Supported networks include Novell, Banyan Vines, Netware, etc.
BGIS Systems 800 NW 33rd Street Pompano Beach, FL 33064 (305) 783-9600, Ext. 318 305-783-9660 (FAX) J. L. Winn

Basis for information: Excerpts from Watson (1995), and telephone communications with supplier.

WinEst

The WinEst program is designed for building col estimators. Quantities are digitized or manually input. The package contains an estimate spreadsheet and database. It can also interface with databases such as Means and Richardson and other software. An environmental database, for estimating remediation costs is also offered.

The software features user-defined codes, unlimited projects and activities per project, and unlimited WBSs per project. Each WBS can have up to 6 levels. Some user-defined cost categories can be added as well. These features can facilitate the inclusion of a full range of environmental costs.

The system was introduced in January, 1993 and is updated three times per year. The latest release is version 2.6, that was released in June, 1995. The system is targeted to the construction industry and architects and engineers. The number of users (not companies) world-wide is approximately 4000. Sample projects and on-line help are included. The supplier offers telephone and consulting support, and on- and off-site training.

The approximate cost of the system is $1990, with $990 for each additional user. The databases range from $495 to $1495. Prices are subject to change.

Minimum Hardware: Will operate on almost any system. Can support all networks that run MS Windows.

Win Estimator
8209 South 222nd Street Suite 6
Kent, WA 98032
(206) 395-3631 or (800) 950-2374 206-395-3634 (FAX) Bill Schaffer

Basis for information: Excerpts from Watson (1995), and telephone communications with supplier in June, 1995.

CHAPTER 4. SCHEDULING AND COST CONTROL SYSTEMS

This chapter describes software systems designed for scheduling and cost control. The organization of this material is as follows. The first section describes systems applicable to High-End Scheduling. Five systems are highlighted : Dekker TRAKKER™ Project Management System, Open Plan Professional, Promavera Project Planner (P3). Project Evaluation System (PXS), and ASA/OR. The second section highlights one program for Low-High End Scheduling (Project Scheduler 6 Version 1.5). The third section highlights one system for Low-End Scheduling (Texim Project for Windows). The fourth section covers systems for Cost/Control Analysis (Cobra 2.0, Cost Time Management (CTM), EasyABC Plus, and Parade). The last section in this chapter provides abbreviated profiles on FOREFRONT and P/X.

HIGH-END SCHEDULING

Dekker TRAKKER™ Project Management System

Prepared by: Dekker Ltd.

Prepared for: Public purchase

Application: Dekker TRAKKER™ is an Activity Based Costing and Performance Management System that integrates project management techniques, including scheduling, process flow, resourcing, costing, earned value, revenue projection, and technical performance measurement, with general accounting into one database. The Activity Based Costing (ABC) techniques can serve to make environmental costs more visible.

Elements of control:

Cost control
Resource control
Estimating control
Schedule control
Scope control
Risk control

Development date and updates: The current version is release 2.9, which was released in August, 1994. Release 3.0 for Windows was available in August, 1995.

Public availability: Available

Purpose and current use: Dekker TRAKKER™ is used in the US and abroad as an enterprise management costing system for commercial and government contracts.

Cost information: The price of the system varies according to the option chosen. The single-user prices for the four options are follows:

Dekker TRAKKER CAMS $2,950
Dekker TRAKKER EMS $6,750
Dekker TRAKKER Plus C/SCSC $9,500
Dekker TRAKKER Tech Pert Meas $12,000

These are prices are approximate.

System requirements: Trakker can be run on single-user or local area networks. It is available for MS-DOS, UNIX, VAX/VMS, and LANS platforms.

Software Summary

Size and complexity of projects: Can be used for various project sizes and complexities.

Other compatible systems offered: Dekker offers four types of software systems to suit individual requirements: Trakker Executive Management System (EMS), Trakker CAMS, Trakker Plus C/SCSC, and Trakker Performance Measurement System.

Non quantifiable information: Note-writing feature.

Trakker is an integrated system for schedule, resource, cost and performance management. It facilitates baseline development, updation, analysis and reporting, and performance measurement. It features Activity Based Costing (ABC) techniques. This summary describes Trakker CAMS, EMS, and Plus C/SCSC.

Trakker Executive Management System (EMS) is designed for business professionals who want to use ABC techniques to track project costs and revenue. Trakker CAMS is designed for project managers who want to implement techniques to monitor project progress in terms of cost, schedule, resource, and earned value. Data from all three systems are compatible.

Trakker CAMS, EMS, and Plus, all feature tools for scheduling (unlimited activities and projects), resource, and cost management (unlimited resources and flexible ceding), earned value (summarized at any level), reports and graphics, and burdening. They are all menu driven and look the same across different platforms. Trakker Plus has additional features for C/SCSC and government reporting. Both EMS and Plus have direct and generic interface with various accounting systems. CAMS does not have this capability but features additional functions for scheduling, and resource and cost management.

Life-cycle stages covered

Raw material acquisition
Manufacturing stage
Use/reuse/maintenance
Recycle/waste management

The system is not designed to consider beyond one life-cycle stage, typically the manufacturing stage.

Type of costs considered

Conventional
Potentially hidden
Contingent
External

The system is designed to include information on conventional costs alone. The activity based costing systems encourage significant breakdown of work activities and the proper allocation of costs, automatically making a number of potentially hidden costs more visible. External or contingent costs could be included/identified through the flexibility of the systems' resource library, direct and indirect rates. If the program is used solely for schedule control, these costs may not bear as much importance.

Method of cost estimation

The systems are not designed for cost estimation.

Generation of financial indicators

Net present value (NPV)

Payback period
Internal rate of return (IRR)
Benefits cost ratio
Other

Ability to Include Environmental Costs

User friendliness and flexibility

The systems can be fully integrated and are menu-driven with similar features across platforms. Trakker can interface directly with company accounting systems. It facilitates government reporting (C/SCSC requirements).

User-support

Dekker organizes workshops on project management and Trakker applications (basic and advanced) and offers telephone and on-line support. No demonstration disks are available. However, Dekker does live on-site software demonstrations upon request.

Limitations

Although the system could potentially include information on all four cost categories, the default formats, sample projects and examples do not specifically explain, include, or refer to all these cost types and categories.

Basis for evaluation: Based on information package received from Dekker in February 1995.

Contact information: Dekker Ltd. 636 East Brier Drive, Suite 260 San Bernadino, CA 92408

909-384-9000 909-889-9163 (FAX) Debra Waiters

Open Plan Professional

Prepared by: Welcom Software Technology (WST)

Prepared for: Public purchase

Application: Scheduling

Elements of control:

Cost control
Resource control
Estimating control
Schedule control
Scope control
Risk control

Development dateand updates: WST was founded in 1984. Open Plan for DOS was released in 1984, and Open Plan Professional was released in December 1994. WST typically offers at least one update every year. Latest release of Open Plan Pro is version 1.1, released in 1995. The latest release of Open Plan for DOS is version 5.1, released in 1994.

Public availability: Available

Purpose and current use: WST has supplied project management software to different industries such as construction, engineering, oil and gas and also to government bodies and universities. WST has a client list of over ISO.

Cost information: The single-user price for Open Plan is $4,200. The single-user prices for Opera and Optic are $2,200 and $995, respectively. The single-user price for Open Plan Pro is $6,000. Prices are subject to change.

System requirements: The minimum system requirements for Open Plan Pro are an IBM (or compatible) 386, 8 Me RAM, DOS 5.0, MS Windows 3.1. MS mouse (or compatible), and any NetBIOS compatible LAN.

The minimum system requirements for Open Plan are an IBM (or compatible) 386, DOS 3.0 (3.1 for LAN version), FoxPro 2.0 or Dbase IV 1.5, and any NetBIOS compatible network.

Software Summary

Size and complexity of projects: Open Plan Pro has been used for projects with budgets of over $2 billion. It has a limited number of WBS levels (20) with a maximum of 30,000 activities per project or subproject. It allows up to 99 user-defined hierarchical or flat code fields. Advanced users can write data-management routines in FoxPro (Open Plan for DOS), allowing incorporation of project complexity (within limits). Open Plan Pro allows for customization through Xbase or Visual Basic. Open Plan is a high-end offering and would not be an economical option for small projects alone.

Other compatible systems offered: WST also offers Opera, an Open Plan extension, for risk analysis and Optic, a project information manager for tracking documents. Open Plan can be used along with other WST software: Texim Project and Cobra (for cost management). Texim Project is currently offered as the low-end project management tool.

Non quantifiable information: Annotation capability.

The system is offered in two forms: the Windows-based Open Plan Professional and the FoxPro-based Open Plan. Open Plan and Open Plan Pro provide advanced functions for project management, mainly scheduling (many more features than Texim). The two share many features and Open Plan Pro has some added features such as the Project Management Director and additional customization features. Open Plan may also have certain features that are not present in Open Plan Pro such as ADM networking and the ability to support compound activities. Features of Open Plan Pro include the following:

· **Project Planning:** PDM networks, up to 30,000 activities per subproject with unlimited projects, 99 user- defined hierarchical or flat code fields.
· **Resource Scheduling:** resource breakdown structure (organization by department, pool, or skill); alternate resource scheduling; resource profiling supports seven user-defined spread curves for loading:user-defined priority rules, definition of resources as normal, consumable, or perishable.
· **Cost Control:** activity costs (budget, actual and performed), resource costs, resource cost escalation, schedule and cost variances.
· **Multi-Project Scheduling:** resource pool sharing and resource leveling across multiple projects, project priorities, results ran-back to project or subproject level.

What-if and cross-project analysis allow users to investigate multiple project scenarios. Data organization is made easier through drag and drop features. Open Plan Pro incorporates risk analysis based on Monte Carlo simulation, thus helping users evaluate uncertainties based on cost and schedule. Views and reports include X/Y charts, histograms, bar charts, networks, breakdown structures, and outlines.

Life-cycle stages covered

Raw material acquisition
Manufacturing stage
Use/reuse/maintenance
Recycle/waste management

Open Plan is not designed to consider more than one life-cycle stage, which is typically the manufacturing stage of the life cycle. Like most scheduling products, the primary area of interest is within the confines of the project start and finish points. However, even this can include the waste management and the use/reuse/maintenance stages, depending upon the time-scale of the project and users' perspective.

Type of costs considered

Conventional
Potentially hidden
Contingent
External

The system has been designed to consider only conventional costs. However, the flexibility in defining WBS structures and activities allows most users to feed in comprehensive cost data if required. Additional cost categories and other such customization cannot be accomplished through menu commands, but can be incorporated through the database.

Method of cost estimation

The system is not designed for cost estimation. However, cost estimates/forecasts can be created in Open Plan and Open Plan Pro.

Generation of financial indicators

Net present value (NPV)
Payback period
Internal rate of return (IRR)
Benefits cost ratio
Other

The Project Management Director, a module in Open Plan Professional, can perform cost-benefit analyses and provide summary cost estimates. No other financial indicators are generated.

Ability to Include Environmental Costs

User friendliness and flexibility

Open Plan Pro has a Windows interface with interactive views and graphical data entry. It features a context-specific help function (customizable in Open Plan). The system can support electronic mail (MS Mail and Lotus cc:Mail), which can assist in data exchange and updating. The products can be used in a multi-user environment with simultaneous updating and file and record-locking capabilities. Open Plan supports multiple projects with up to 20 WBS levels per project. Open Plan Pro and Open Plan (for DOS) can export and import data from Primavera, ASCII text files, and MS Project. Open Plan can also exchange data in the U.S. Army Corps of Engineers Standard Data Exchange Format. This capability does not currently exist in Open Plan Pro.

Open Plan Pro includes sample projects (e.g., an environmental project) that can guide users and give examples of the types of fields and data to be entered. Open Plan Professional has default templates that detail generally accepted project management practices and help users forecast: track; and control project scope, time, cost, human resources, communications, quality, risk, and contract/procurement management. These templates can be used as is or customized to meet specific project requirements. Open Plan Pro includes features for customizing reports. Both Open Plan Pro and Open Plan can be customized, although in different ways. Users cannot make custom changes through a menu item. They would need to do that through the relational database or through external languages (such as Visual Basic).

User-support

Open Plan Pro includes a computer-based training facility that helps reduce training costs for new users. WST offers training, consulting, and technical support to assist users in implementing their projects. They provide demonstration software, press reviews, and references upon request. Maintenance support is provided free of cost for the first year with the option for annual renewal at a cost of 20% of the (current) retail price thereafter. The support includes technical support, enhancements, and updates and access to the bulletin board where users can exchange information. Slide-show demo disks are available for Open Plan Pro.

Limitations

Open Plan does not provide users with information about including all cost categories. Not unlike other software systems, it is designed to consider only one life-cycle stage. Customization is not as straightforward in Open Plan. However, users can integrate this system with another WST product Cobra (profiled elsewhere in the report), which can afford much greater flexibility in defining costs and cost management.

Basis for evaluation: Based on information package received from WST in March 1995. Additional information was obtained through telephone communications with WST representatives in April 1995.

Contact information: Welcom Software Technology (WST) 15995 North Barkers Landing Road Suite 275 Houston, TX 77079-2494 Joyce Janda (National Sales Manager) 713-558-0514 713-584-7828 (FAX) 800-247-4WST

Primavera Project Planner (P3)

Prepared by: Primavera Systems, Inc.

Prepared for: Public purchase

Application: Primavera Project Planner (P3) for Windows (or DOS) for project management and
scheduling.

Elements of control:

Cost control
Resource control
Estimating control
Schedule control
Scope control
Risk control

Development date and updates: DOS version was first released in 1983. Windows version was released in May, 1994. The current DOS release is version 5.1, released in May, 1993. The current Windows release is version 1.1, released in April, 1995.

Public availability: Available

Purpose and current use: Cross industry. P3 currently has approximately 40,000 users (not sites).

Cost information: The single-user price for P3 is $4,000.

System requirements: P3 can be used in the Windows or DOS format and on a network. Other system requirements: 486 PC 6 Me RAM · 25 MB disk space

Software Summary

Size and complexity of projects: Suitable for large projects. Can support unlimited projects, with up to 100,000 activities.

Other compatible systems offered: Primavera's line of project management software includes Parade for performance measurement and Monte Carlo for risk analysis, which are discussed elsewhere in the manual. It also includes Expedition Contract Control software for managing engineering and construction projects and SureTrak Project Scheduler for planning and presenting small- to medium-sized projects.

Non quantifiable information: Project logs. Email data exchange P3 has been designed for project scheduling and resource and cost control. It uses the Precedence Diagramming Method (PDM) for scheduling, and 14 types of schedule constraints and four different relationships can be used. Some features of P3 are as follows:

· What-if and earned value analysis-detailed notes on each activity. Filtering option or the what-if analysis make analyzing different management scenarios and strategies possible.
· Multiple forecasting techniques.
· Complete Customization of reports and graphics with over 200 predefined reports and graphics.

· 24 activity codes for selection, sequencing, and filtering data. Filtering allows isolation of activities or segments of the project that need analysis.
· Up to 16 custom data items to organize and track details.

Life-cycle stages covered

Raw material acquisition
Manufacturing stage
Use/reuse/maintenance
Recycle/waste management

P3 is not designed to consider more than one life-cycle stage, which would typically be the manufacturing stage of the life cycle. That is, it does not have a built-in mechanism for differentiating between the various life-cycle stages. However, life-cycle stages could be distinguished within P3 in different ways, such as through the WBS levels.

Type of costs considered

Conventional
Potentially hidden
Contingent
External

P3 has been designed to consider only conventional costs. However information on other types of costs could be incorporated into P3 by adding additional cost categories, through activity information, and through ceding (for example, codes can be assigned to work packages). However, activity codes for filtering data in P3 are limited (24) and this could become a constraint if that option is employed.

Method of cost estimation

The system is not designed for cost estimation.

Generation of financial indicators

Net present value (NPV)
Payback period
Internal rate of return (IRR)
Benefits cost ratio
Other

Ability to Include Environmental Costs

User friendliness and flexibility

P3 can be extended for direct interface with other Primavera software such as Parade and Monte Carlo. It is also possible to exchange data with other Windows products, dBASE, Lotus, and U.S. Army Corps of Engineers exchange format. It has a batch processor that allows import and export of ASCII data. P3 has many features such as the ability to handle multiple projects in a multi-user environment and to generate specialized reports and graphics.

P3 permits unlimited projects (master projects, subprojects, target plans), up to 100,000 activities per project, and unlimited cost accounts per project. P3 allows activity information to be organized and displayed by both resource and time period An unlimited number of resources can be used per project and per activity. Variable resource prices and availability can be entered and information on variable resource prices and availability can be entered. It also allows unlimited cost accounts per project. There is an autocost feature for user-defined rules for planning and updating costs. P3 can analyze the data from any angle to identify problems and isolate areas of concern or corrective actions required. The online Help makes using the system easier for inexperienced users. There is a Global change option for mass changes and updates. It is also possible to use parts or fragments of earlier projects (Fragnets) to build a new project.

The above mentioned features allow some flexibility in entering cost information. Thus users may be able to create a separate subprojects or activities within projects (e.g., "waste management" or "external costs").

User-support

Telephone hotline support and bulletin board. Working demonstration disks for P3 are available, that are limited by the maximum number of activities supported (60).

Limitations

P3 has been designed to consider only conventional costs. The system does not include information on other cost types in default information or sample projects.

Basis for evaluation; Based on information received from Primavera Systems in February 1995.

Contact information: Primavera Systems Inc.
Two Gala Plaza Gala Cynwyd, PA 19004
Nicole Stephana
610-667-8600 610-667-7894 (FAX)

Project Execution System (PXS)

Prepared by: Grantlun Corporation, "Construction information solutions"

Prepared for: Public purchase

Application: Project Execution System (PXS) for project management.

Elements of control:

Cost control
Resource control
Estimating control
Schedule control
Scope control
Risk control

Development date and updates: Version 2.0 of PXS is due to be released in July, 1995. This system will be updated at least once a year.

Public availability: Available

Purpose and current use: Hard Dollar systems are designed only for the construction industry, for heavy/highway bidding and building. PXS can be used for a variety of industries and projects (that typically involve procedures similar to those in construction projects).

Cost information: The price of PXS has not yet been fixed. It is anticipated tobe about $8,000 to 9,000 (single user).

System requirements: PXS requires an IBM compatible 486-25 MHz or faster PC, with 4 MB RAM, 80 MB available hard disk space and a VGA monitor. It can be run as a stand-alone system or on a network.

Software Summary

Size and complexity of projects: The system is not modular in structure and therefore may not be suitable for small projects. PXS has functions to handle project complexities within certain limitations.

Other compatible systems offered: Hard Dollar EOS

Non quantifiable information: Notes can be added alongside any job. PXS is linked with Hard Dollar EOS, which manages the estimating and bidding side, while PXS handles the building (project management and control). PXS gives users many tools besides the scheduling and accounting systems. PXS is available in multiple modules that include the following:

· **Analyze module:** establishes the work that needs to be done. Preliminary WBS trees can be set up with data that are entered or imported.
· **Plan module:** establishes resource requirements and duration of work. Description of critical activities, work structure, and resources.
· **Schedule module:** manages the job schedule and registers progress. Automatic creation of a preliminary schedule. Schedules can be created at any level and current and archive schedules can be maintained for claim and change order support.
· **Procure module:** manages the procurement of services and

resources. Vendor and subcontractor management, including user-defined procurement constraints and logs of payments and work progress.

· **Dispatch module:** manages the assignment of work plan activities to specified resources. Daily dispatch orders that can be sorted by resource or work activity.

· **Administer module:** manages the administration of changes, contract documents and cash flow. Includes an integrated word processor. Features for administering project budgets, cash flows, projections, and changes. Tools to determine change-order costs and impacts of schedule changes.

· **Document module:** manages the documentation of job activity. Includes maintenance and job correspondence, transmittals, minutes, phone logs, and a daily diary.

· **Monitor module:** monitors and analyses job progress. Tools for tracking and monitoring progress, including costs, payments, and procurements.

PXS can also prepare detailed budgets, resource utilization reports, and cash flow schedules with projected cost/revenue curves. Each work item can be broken down into five levels of detail with unlimited activities at each level. Users can maintain a database of current labor, equipment, and other resource rates.

Life-cycle stages covered

Raw material acquisition
Manufacturing stage
Use/reuse/maintenance
Recycle/waste management

The system is designed to typically consider only the manufacturing stage of the life cycle.

Type of costs considered

Conventional
Potentially hidden
Contingent
External

The system is designed to consider only conventional costs in various levels of detail. Information on other cost types could be entered into the system through activity information (unlimited activities per level).

Method of cost estimation

The system itself is not designed for estimating costs, but cost estimating data can be input from Hard Dollar EOS (described in the cost estimating tools section).

Generation of financial indicators

Net present value (NPV)
Payback period
Internal rate of return (IRR)
Benefits cost ratio
Other

Ability to Include Environmental Costs

User friendliness and flexibility

EOS and PXS can share data and integrate the bidding and building process. Hard Dollar systems have been designed to allow nonexperts to use them with ease. A unique feature of PXS is the voice activation option. PXS has functions that many other scheduling systems do not have, such as integrated spreadsheets and modules for dealing with vendors and subcontractors. Work activities can be imported from spreadsheets, estimating systems, digitizers, text files, and electronic files from public agencies. PXS can be run on almost any operating system, including MS Windows, DOS, UNIX, and Macs. It can also be run on portable DOS-based computers. The system has provided users with options for customizing reports and applications.

User-support

Hard Dollar customers can take advantage of training sessions and 24-hour support (as part of the annual renewal fee) offered by the company. Hard Dollar systems have aeon designed to allow nonexperts to use them with ease. The price includes free support and maintenance for one year, which includes toll-free support and all updates and enhancements. After the first year, users have the option to renew the maintenance contract. The price of this varies. Limited time demonstration software can be provided to potential users.

Limitations

The system has been introduced only recently, although the vendors have the experience of handling Hard Dollar EOS. Since the system has been designed specifically for the construction industry, it focuses on communications and contract management (vendors and subcontractors) in a level of detail that may not be required in many other projects.

Basis for evaluation: Based on information package received from Grantlun Corp. in November 1994. Additional information obtained through telephone communications with Grantlun representatives in

March 1995.

Contact information: Grantlun Corporation "Construction Information Solutions" 2055 E. Centennial Circle, Suite 107 Tempo, AZ 85284
Ryan Brady
800-637-7496 602-897-6939 (FAX)

SAS/OR

Prepared by: SAS Institute Inc.

Prepared for: Public purchase
Application: SAS/OR, an integrated component of the SAS system, provides tools for project management, decision analysis, and mathematical programming. A host of other SAS products can be used in conjunction with the SAS/OR software. Project planning, management, scheduling, and tracking can be accomplished using a single integrated system. Complementing the SAS/OR project management tools is the PROJMAN menu system, a customizable and extendable point-and-click interface for scheduling projects and producing charts and reports.

Elements of control:

Cost control
Resource control
Estimating control
Schedule control
Scope control
Risk control

Development date and updates: SAS/OR software was released in

1983. Every new release for the SAS system (on an average once in two years) will add features to the SAS/OR software. The next release is expected in the end of lass/the beginning of 1996. The current release is version 6.10(1994).

Public availability: Available

Purpose and current use: SAS Institute was established in 1976 to develop and market base SAS software. Their flagship product was originally developed to analyze agricultural data on IBM mainframes at North Carolina State University. In the 1980's the SAS system was completely rewritten in C to incorporate a MultiVendor Architecture (MVA).

Cost information: License fees and prices will vary based on options chosen, such as platforms and number of users. Users need to pay an annual license fee to be able to use the systems. The license fee for SAS/OR for UNIX and PCs can start from 770 for the first copy (single user) with $360 for annual renewal. Different charges are applied for additional users, for example, for 10 copies the first year fees is $3770 and the renewal fee is $1890. The price decreases as the number of users increase. These are 1995 prices, and are subject to change.

System requirements: The MVA architecture makes the 90% of the SAS system code portable across all hardware platforms (at the most 10% needs to be rewritten). The SAS system has been released for various platforms: MVS. CMS, VSE, Open VMS for VAX and AXP, OS/2 2.0. Windows 3.1, Windows NT, MAC, DG/UX. SUN OS, RISC/OS, OSF/1. MIPS. ABI. INTEL ABI, AIX, HP-UX, RISC/ULTRIX, Solaris, and ConvexOS. The specific system requirements will vary depending on the platform chosen. The program can be made network compatible, supporting various networks including NETBIOS, TELNET, APPC for Windows, and others.

The graphics devices may require the availability of SAS/GRAPH software (for high resolution graphics) since SAS/OR uses fonts and device drivers provided with the SAS/GRAPH software.

Software Summary

Size and complexity of projects: An unlimited number of user-defined identification variables can be defined to capture WBS structure and sub-project information. There is full control over logical precedence relationships. Scheduling can also be in seconds, minutes, or unspecified (relative time units). Project calendars are completely user-defined. An unlimited number of project resources and resources per activity can be defined. The system can also handle multiple projects. All these features and others provide users with a lot of flexibility in customizing projects and incorporating project complexities, and project sizes.

Other compatible systems offered: SAS Institute develops a number of different products that can be used along with SAS/OR. These include decision analysis tools, econometric modeling, quality control, statistical packages, graphics, and others. SAS/OR software is a fully integrated component of the SAS system.

Non quantifiable information: The current release of SAS/OR does not have a separate 'notes' field, but does allow annotation on charts and graphs. The next release will have a space for adding unlimited project notes.

SAS/OR forms an integrated component of the SAS system and includes several basic types of programs. These are procedures for scheduling and project management, for decision analysis, and for general mathematical and network flow programming. The project management capabilities can be performed through a menu system called PROJMAN. The software includes procedures for exploring models of distribution networks, production systems, resource allocation problems, and scheduling problems using tools of operations research. Features of SAS/OR software include:

· **Editing project data:** Project data includes activity, logic, resource (and costs), and calendar data. The software includes features such as maintaining activity information, activity successors and lags, multiple activity calendars, and specification of resource availability and requirements

· **Scheduling:** interactive control of the scheduling algorithm, limited resource availability, resource priorities, substitutable resources, multiple precedence relationships, constraints, monitoring, and additional options including activity splitting and infeasibility diagnostics.

- CPM procedure for scheduling projects using the Critical Path Method
- GANTT charts for monitoring project progress
- NETWORK diagrams with graphical features for visualizing structural relationships, WBS, and organizational charts

· **Decision analysis:** interactive procedure that interprets decision problems, finds optimal decisions, and plots the decision tree.

· **Mathematical programming:** solution of linear, integer, and mixed integer problems and non-linear optimization.

· **Reporting:** report-writing capabilities in the SAS system allow users to easily transform output into a variety of customized reports and graphics. Simple or sophisticated tables, plots, charts, maps, network diagrams, gantt charts, and more, can be generated without leaving the SAS system.

Multiple projects can be scheduled with a common pool of resources. Cost data across multiple projects can be summarized and analyzed, and earned value analysis can be performed. A virtually unlimited number of what-if solutions can be evaluated.

Life-cycle stages covered

Raw material acquisition
Manufacturing stage
Use/reuse/maintenance
Recycle/waste management

The system is not designed for life-cycle costing. However, required life-cycle information can be incorporated into the system using the SAS/OR components and other applications. The system would typically be used to consider portions of the manufacturing stage of the life-cycle.

Type of costs considered

Conventional
Potentially hidden
Contingent
External

The SAS system treats costs as a resource. Applications can be developed to handle cost management. As is, the system does not include features for including hidden, contingent, or external costs. However, these costs could be included with appropriate Customization of the system, such as through the creation of additional cost categories, user-defined codes, and cost reporting by 'type' of money.

Method of cost estimation

The system is not designed for cost estimation. However, applications for performing various types of cost estimation could be created using the SAS system tools.

Generation of financial indicators

Net present value (NPV)
Payback period
Internal rate of return (IRR)
Benefits cost ratio
Other

The SAS system includes procedures for computing NPVs and IRR for uniform payments. Programs for computing payback periods, benefits cost ratio, and other indicators could be created using the SAS system. The SAS system enables users to outline the basics of a project, estimate the profit/loss characteristics and perform an initial cost analysis to decide if it is really worth executing.

Ability to Include Environmental Costs

User friendliness and flexibility

The SAS system is a 4GL. The PROJMAN menu system that complements the SAS/OR software is menu driven and users do not need to learn the SAS language in order to use it. Many charts and diagrams can be produced in either line printer, full screen, or high resolution graphics form. Customized reports can also be prepared. Screen options can be customized, including the modification of global menu parameters that control default scheduling options. Sample protect data sets can be created, and sub-projects and/or individual project data sets can be copied, deleted, or combined. External existing SAS data sets can be imported into PROJMAN. The CPM, NETDRAW, and GANTT procedures integrate with the SAS system and allow users to develop a customized project management system. As with other SAS programs, users can access the codes in PROJMAN to make any modifications and updates. Thus items that are not currently in the menu can be added in this manner. However, this will require knowledge of the SAS language and procedures. The next release of PROJMAN will make the system completely different. Features will be added to make the system more interactive and more object-oriented.

The SAS system can access, read, and write data in a variety of different data structures and file types, including relational database management systems, hierarchical database management systems, data gateways and applications programming interfaces, operating system files, various external file formats, and SAS data sets. Also, the SAS system is portable across computing environments, which means that it will look and function in the same manner on all platforms.

Users can choose to include a variety of different ready-to-use programs of procedures for various applications, from general purpose data processing to specialized analyses. These include statistical and mathematical analyses, financial reporting, market analysis, etc.

User-support

The system includes an built-in tutorial that provides information on the menu system and basic features of the SAS system. No demonstration disks are provided. However, the supplier offers a free, no risk 30-day trial of the software. The supplier offers a variety of services and support options to users. This includes technical support, training (on- and off-site in various levels and formats), publications (training manuals, guides, newsletters), consulting services, marketing and development partnerships, and membership in national, regional, local, in-house, and special interest SAS user groups.

Limitations

SAS/OR does not have special features or default formats to facilitate the inclusion of environmental cost information in project management. However, the flexibility of the system will allow users to include these features at their discretion. The SAS system itself is not menu driven and requires knowledge of the SAS language for use and manipulation. PROJMAN is menu driven, but in its current format does not include every feature in the form of menu commands.

Basis for evaluation: Based on information received from the SAS Institute in March, 1995.

Contact information: SAS Institute Inc. SAS Campus Drive Gary, NC 27513 919-677-8000 919-677-8123 (FAX)

LOW-HIGH END SCHEDULING

Project Scheduler 6 Version 1.5

Prepared by: Scitor Corporation

Prepared for: Public purchase

Application: Project management.

Elements of control:

Cost control
Resource control
Estimating control
Schedule control
Scope control
Risk control

Development date and updates: The system was introduced in 1983. Project Scheduler was the world's first DOS-based PM application. It was also among the first to allow multi-project management, and incorporate a graphical user-interface into the application. Project scheduler 6 version 1.5 was released in late 1993.

Public availability: Available

Purpose and current use: Scitor Corp. started its operations as a consultant to the Aerospace industry. Project Scheduler has users in both Federal facilities and in private industry, including pharmaceuticals, consulting, insurance, construction, and engineering.

Cost information: $695 for lan-ready individual units or, $2835 for a 5-user license. Prices are subject to change.

System requirements: 386 (or higher) PC / 4 Me RAM / DOS 4.0 (or higher) / Windows 3.1
/ VGA (or higher) resolution monitor / Hard drive with 4 MB available / 3.5" high density (1.44 Me) disk drive

Software Summary

Size and complexity of projects: The system is not modular in structure and is designed for mid to high-end project management. It is limited by the maximum number of activities (20,000) and resources per project (10,000) that it can handle. However, the system includes a number of features typically found in high-end PM systems, including variance analysis and multi- project planning.

Other compatible systems offered: None

Non quantifiable information: Each task (activity) has a field for entering unlimited notes. Parts of these fields can also be displayed on the Gantt chart.

Features of Project Scheduler for Windows include the following.
· multi-project planning and tracking. As many as 10,000 shared resources among an unlimited number of projects.
· CPM, network diagram (PERT chart), custom gantt charts, and user-configurable spreadsheets
· what-if analyses (unlimited undo/redes). PERT analysis mode for calculation of best, worst, and most likely project schedules. Variance analysis and performance measurement.
· data arrangement, sorting, and filtering options with WBS, DBS,

or a Resource Breakdown Structure (RBS). Codes can be automatically derived. The WBS, DBS, and RBS can each have up to 10 levels.

· advanced resource tracking spreadsheet to analyze task assignments on a period by period basis.
· automatic resource leveling (priority codes can be assigned at the task and project level), resource splitting, and automatic conflict resolution and recalculation
· user-defined task, project, resource, and assignment fields. Formula fields for project, task, and assignment levels. Custom calendars
· 40 preconfigured reports with Customization features
· supports the Open Database Connectivity (ODBC) interface standard
· context-sensitive on-line help

Life-cycle stages covered

Raw material acquisition
Manufacturing stage
Use/reuse/maintenance
Recycle/waste management

The system is designed to typically consider only the manufacturing stage of the life cycle. The WBS allows tasks to be arranged by life-cycle phases. Up to 10 WBS levels can be defined. However, the only life-cycle stages of interest would be those occurring during the span of the project. The system is not meant for life-cycle cost analysis.

Type of costs considered

Conventional
Potentially hidden

Contingent
External

The system is designed to consider only conventional costs in various levels of detail. Other cost categories could be considered through adding appropriate activities. These activities could be isolated or grouped using the ceding mechanism. There are 10 different cost fields, in which 7 are user-defined. This will allow users to detail additional cost categories, including hidden, contingent, and external costs.

The resource table allows users to enter a standard rate along with five alternate costs for each resource. This may be a way to analyze the effects of factoring in hidden and external costs into resource costs. Up to 10 inflation schedules can be saved. Overhead costs can be factored in using a custom formula that multiplies standard data fields by the values assigned by the formula. User-defined mathematical formulas can be established at the project, task, and assignment level to automatically reflect the appropriate markup, discount, or custom calculations. This can be utilized to factor in estimates of contingent or liability costs.

Method of cost estimation

The system is not designed for cost estimation.

Generation of financial indicators

Net present value (NPV)
Payback period
Internal rate of return (IRR)
Benefits cost ratio
Other

Project Scheduler does not have any such financial functions, but it can interface with other software systems (e.g. MS Excel) that do offer these functions.

Ability to Include Environmental Costs

User friendliness and flexibility

The system imports and exports data using ASCII files or via its ODBC interface. It also reads files in Project Scheduler 5, Microsoft Project extension (.MPX), and comma-separated variable (.CSV) format. The ODBC support allows users to export and import specific project data fields to other ODBC capable networking applications such as spreadsheets and relational databases.

The multi-project planning, 10 levels for WBS (also OBS and RBS), resource tables, and the user-defined cost categories afford users with the flexibility to enter life-cycle and environmental cost information in a number of different ways.

User-support

Project Scheduler comes with a tutorial manual, a user's guide, an addendum covering changes in Version 1.5, and a reference manual. The program includes on-line help, sample project files, and indexes. Scitor offers unlimited free (but not toll-free) telephone support. They also support a CompuServe forum. Consulting and training services are also offered at extra charges.

Limitations

Project Scheduler does not provide any step-by-step context-sensitive guidance with building project plans. It does not include default cost categories for covering potentially hidden, contingent, and less tangible costs.

Basis for evaluation: Based on information package received from Scitor Corp. in
April, 1995.

Contact information: Scitor Corporation 393 Vintage Park Drive Suite 140 Foster City, CA 94404
Rick Garner (Sales team)
415-570-7700 415-462-4301 (FAX)

LOW-END SCHEDULING

Texim Project for Windows

Prepared by: Welcorn Software Technology (WST)

Prepared for: Public purchase

Application: Low-end project management. Users can design their projects to incorporate environmental costs. However, no additional cost categories can be defined in Texim.

Elements of control:

Cost control
Resource control
Estimating control
Schedule control
Scope control
Risk control

Development date and updates: Texim was earlier released by another company, which was purchased by WST. The Windows version was released -1991. WST features at least one update per year. Texim is going to be replaced within about a year with another low-end project management system. Open Plan Professional-Desktop version. Texim version 2.0 was released in 1994.

Public availability: Available

Purpose and current use: WST has supplied project management software to different industries such as construction, engineering, oil and gas and also to government bodies and universities. WST has a client list of over 150.

Cost information: Texim has a single-user price of $995.

System requirements: The minimum system requirements are a 386 SX CPU, mouse or other pointing device, and any NETBIOS compatible network. For MS Windows 3.1 in enhanced mode, the requirements include 4 MB memory, WIN 32s included. For MS NT, the requirement is 8 MB memory.

Software Summary

Size and complexity of projects: Small to large projects with limited complexity.

Other compatible systems offered: Other products supplied by WST include Open Plan (DOS) and Open Plan Professional for high-end project management, Texim Project for low-end project management, Opera for risk analysis, and Optic for information management.

Non quantifiable information: Note writing capability (limited).

Texim Project is designed for lower-end project management and includes features such as time analysis, resource scheduling, cost management, and risk evaluation. Texim is built on Windows and Windows NT. WST suggests that Texim Project may also be used as a Windows front-end to feed data into Open Plan. The system can also be used for developing proposals, after which detailed development can be done in Open Plan. Some features of Texim project include the following:

· **Scheduling Tools:** unlimited projects per file, 25 WBS levels, 1000 activities per project plan, 1,000 activities per subproject, 20 user-defined code-fields, CPM scheduling, resource-limited scheduling, resource leveling, project risk analysis (Monte Carlo method).
· **Resource Tools:** 25 OBS levels, unlimited pools, 1,000 resources per pool and resources per activity, crucial and noncrucial assignments, partial allocations, heuristics.
· **Cost Tools:** WBS and OBS cost accounting, fixed and variable activity cost, fixed and periodic resource cost and fixed, variable or percent, project, and pool overhead.
· **Tracking Tools:** Flexible activity and resource status determination, up to 60 baselines per project, earned value, cost estimates at completion, performance profiles.
Texim can generate graphs such as resource and pool histograms and cost and schedule index curves. What-if analysis can help users analyze alternative project scenarios.

Life-cycle stages covered

Raw material acquisition
Manufacturing stage
Use/reuse/maintenance
Recycle/waste management

The system has not been designed to consider more than one life-cycle stage, which typically would be the manufacturing stage of the life cycle. The limited number of activities may make incorporating life-cycle cost information difficult. The WBS levels can also be utilized to input life-cycle information.

Type of costs considered

Conventional
Potentially hidden
Contingent
External

Texim project has been designed to consider only conventional costs. Users can design their projects to incorporate other costs as well. However, no additional cost categories can be defined in Texim. Costs will need to be entered by adding activities. Data sorting and filtering options may also be used for segregating information on different types of costs.

Method of cost estimation

The system is not meant for cost estimation.

Generation of financial indicators

Net present value (NPV)
Payback period
Internal rate of return (IRR)
Benefits cost ratio
Other

Ability to Include Environmental Costs

User friendliness and flexibility

Texim has an on-line help system and a menu-driven Windows environment, which make tasks simpler for inexperienced users. It is possible to customize screen-views and reports. Resources and work can be scheduled using graphical year or month calendars. Texim also features automatic cost and schedule aggregation. Data can be sorted, filtered, or highlighted according to users' requirements and data roll-up options help manage multiple projects.

User-support

WST offers training, consulting, and technical support to assist users in implementing their projects. They provide press reviews, and references upon specific request. Maintenance support is provided free of cost for the first year with the option for annual renewal at a cost of 20% of the (current) retail price thereafter. The support includes technical support, enhancements and updates, and access to the bulletin board where users can exchange information. No demo software of offered for Texim.

Limitations

Texim is designed for low-end project management but features seamless data integration with Open Plan, which provides an option for future expansions or the need for advanced features. However, WST does not offer modular options, which afford greater flexibility and can be more cost-effective for some users. Because Texim is designed for low-end project management, it does not come with the same functionality and flexibility of the higher-end tools. Texim is not built around a relational database, and the proprietary nature of the system limits Customization by users. There is no limit on the time horizon of projects, but the number of periods may be limited. Thus, the limit depends on the manner of scheduling (by day, month, etc.).

WST will be replacing Texim Project in about a year with another product called Open Plan Professional-Desktop version. This product will have the same source code as Open Plan Professional and limited functionality.

Basis for evaluation: Based on information received from WST in February 1995.

Contact information: Welcom Software Technology (WST) 15995 North Barkers Landing Road Suite 275 Houston, TX 77079-2494
Joyce Janda (National Sales Manager)
713-558-0514 713-584-7828 (FAX) 800-247-4WST

COST/CONTROL ANALYSIS

Cobra 2.0

Prepared by: Welcom Software Technology (WST)

Prepared for: Public purchase

Application: Cost management. Cobra has been designed to consider only conventional costs, but could accommodate the inclusion of other cost categories as well. It is not designed for life-cycle cost analysis.

Elements of control:

Cost control

Resource control
Estimating control
Schedule control
Scope control
Risk control

Development date and updates: WST founded in 1984. Cobra was released in 1989. Cobra version 2.1 released in 1994. Updates are typically one (or more) per year.

Public availability: Available

Purpose and current use: WST has supplied project management and cost management software to different industries such as construction, engineering, oil and gas and also to government bodies and universities. WST has a client list of over ISO. Cobra is also marketed abroad in countries in Europe, South America, Asia, and Africa.

Cost information: The single-user price is $8,500.

System requirements: The minimum system requirements are an IBM 386 or compatible computer, MS DOS 3.1,1 ME RAM, and FoxPro 2.0. Cobra runs on most PC network operating systems and allows concurrent data access in a multi-user environment.

Software Summary

Size and complexity of projects: Typically used in higher-end systems.

Other compatible systems offered: Other products supplied by WST include Open Plan and Open Plan Pro for high-end project management, Texim Project for low-end project management. Opera for risk analysis, and Optic for information management.

Non quantifiable information: Some sections include a note writing capability.

Cobra is designed for cost and performance measurement and provides budget, actual cost, and earned value information. It features trend, summary, exception reporting, and cash flow management. No financial indicators are calculated. However, the program does calculate earned value and can generate various graphs that depict current performance and future performance based on estimated actuals. What-if analysis can be based on work package, cost account, resource element, or hierarchical structure elements. Changes may be applied as percentages, target amounts, or change amounts. There is no limit on the length of the projects. The periods that can be reported or printed at one time may be limited. Cobra has been designed to integrate with users' existing schedule and financial systems.

Life-cycle stages covered

Raw material acquisition
Manufacturing stage
Use/reuse/maintenance
Recycle/waste management

The system is not designed to consider more than one life-cycle stage, and it would typically consider the manufacturing stage of the life cycle. However, users can apply different methods of including life-cycle information. For instance, they can define life-cycle stages in a WBS structure (up to 20 levels available) or through code files (defining work packages and assigning specific codes to them). The former method will probably work better with sequential information. The flexibility of Cobra allows users to sort and report information on particular WBS levels or work packages.

Type of costs considered

Conventional
Potentially hidden
Contingent
External

Cobra has been designed to consider only conventional costs. However, the flexibility of the system allows users to enter any cost information they require. Cobra allows user-defined cost classes and an unlimited number of rate and rate tables (for different or combined cost classes). It also allows a Resource Breakdown structure. The number of cost classes is limited to 16 to 20 and it is advisable to restrict their use for things that need different rate tables. Users may be able to enter information on different cost categories by designating separate resource categories or by defining sorting codes. Cost categories (e.g., external costs) do not qualify per se as being resources as they are defined (a source of project support assigned to activities), but users may be able to perform the desired functions by assigning them in this manner. The data-filtering option can be used to exclude or include different cost categories defined in this manner.

Method of cost estimation

The system is not designed for cost estimation.

Generation of financial indicators

Net present value (NPV)
Payback period
Internal rate of return (IRR)
Benefits cost ratio
Other

The system does not generate financial indicators within the system, but integrating Cobra with financial systems that may perform these functions is possible.

Ability to Include Environmental Costs

User friendliness and flexibility

WST offers Cobra as a system for enabling complete cost and schedule integration with their scheduling software products. Open Plan and Open Plan Professional. However, Cobra can import from CPM software (via ASCII files) and pricing software as well. Cobra supports ASCII or dBase transaction files for actual cost collection. It can export data to Lotus 123 and Excel.

Besides having a set of built-in reports, the software supports user-defined report conditions and custom reporting. Cobra adheres to government-reporting criteria for contractors involved in defense, aerospace or energy contracts in the U.S and elsewhere. The attributes include a context-specific help function, pun-down menus, and user-defined prompts and responses.

The package allows up to 20 WBS levels, with up to 190 children per element in a level. Also, users have the ability to define cost classes and rates and assign resources in various ways. Users can maintain up to four separate forecasts per project and these may be entered manually as bottom-up estimates or generated automatically via efficiency factors. The program also supports subcontract management. This flexibility can support various kinds of projects, with unique budget and cost information, and the evaluation of different hypothetical scenarios.

User-support

WST offers training, consulting, and technical support to assist users in implementing their projects. They provide press reviews, and references upon specific request. Maintenance support is provided free of cost for the first year, with the option for annual renewal at a cost of 20% of the (current) retail price thereafter. The support includes technical support, enhancements and updates, and access to the bulletin board where users can exchange information. No demonstration software is available for Cobra.

Limitations

Cobra does not provide users with information about including all cost categories. Not unlike other software systems, it is designed to consider only one life-cycle stage. However, as mentioned previously, its flexibility may allow users to include this life-cycle information. Although an unlimited number of resources can be defined in Cobra, the system can slow down if too many are listed. However, the functionality afforded by FoxPro allows a lot more data than, for instance, a Windows product could support (without substantial reductions in speed). Cobra can be used on low-end systems but may be too expensive for users who need it for this purpose alone.

Basis for evaluation: Based on telephone communications and information package received from WST in March 1995.

Contact information: Commercially available
Welcom Software Technology (WST) 15995 North Barkers Landing Road Suite 275 Houston, TX 77079-2494
Joyce Janda (National Sales Manager) Ruth Ann Schulte (Technical)
713-558-0514 713-584-7828 (FAX) 800-247-4WST

Cost Time Management (CTM)

Prepared by: Westinghouse Electric Corporation

Prepared for: Westinghouse Electric Corporation

Application: Diagnostic tool for analyzing project time and costs. Since this has potential to be used
in any application, it can be used for analyzing environmental projects as well. Environmental costs, if available, can be included in any analysis.

Elements of control:

Cost control
Resource control
Estimating control

Schedule control
Scope control
Risk control

Development date and updates: Information not available

Public availability: Available in conjunction with consulting services from the Productivity and Quality Center at WEC.

Purpose and current use: Currently being used in various Westinghouse projects and facilities. It is suitable for applying to a wide range of problems and manufacturing processes.

Cost information: Since the system is offered only in conjunction with consulting services, the price will vary.

System requirements: MS Excel for Windows.

Software Summary

Size and complexity of projects: May be a difficult task to incorporate project complexity, because the project needs to be broken down into manageable components. It can conceptually handle all project sizes.

Other compatible systems offered: None

Non quantifiable information: None

CTM was developed as a diagnostic technique to identify opportunities to reduce cycle time and costs associated with environmental restoration and waste management. CTM uses a cost-time profile of a process to serve as a basis for analysis. A cost-time profile is a graphic representation of the accumulation of cost over time as a product or service passes through its entire production cycle. CTM has been used at Westinghouse facilities to analyze removal action processes, minimization of hazardous waste streams, RCRA permit processes, construction changes, and CERCLA soil sampling and analysis processes. They report that the technique has been successfully applied and has resulted in savings in time and money.

A cost-time profile illustrates the way costs build up over time as a product or service passes through its entire cycle in a business operation. In the profile, vertical lines represent the cost of materials and services. Horizontal lines are wait times, when no work is being done. Finally, the diagonal lines represent work in dollars per hour. The slope of the line depends on the workers' pay rates and how long the work takes. The profile may reveal hidden relationships or improvement opportunities that may otherwise be missed. The idea is to find ways to shrink the profile by identifying solutions that are the most appropriate and that have the greatest impact. A decrease in the height of the profile comes from reducing costs (increased operating profit), and reducing the time axis results in improved quality (and customer service).

Life-cycle stages covered

Raw material acquisition
Manufacturing stage
Use/reuse/maintenance
Recycle/waste management

The system is not designed to analyze more than one life-cycle stage, and it would typically be the manufacturing or use/reuse stages of the life cycle.

If users have cost and time information related to other life-cycle stages, they use this software as an analysis tool. The information can be analyzed in stages or together. However, this information (especially time factors) can be difficult to obtain.

Type of costs considered

Conventional
Potentially hidden
Contingent
External

The system has probably been defined to analyze usual and hidden costs (judging from the projects it has been used for). However, because costs are not categorized in any way, users can include any kind of costs in their analysis. However, the literature does not say how any useful analysis can be accomplished using these other categories. For example, because a company does not directly incur external costs, users may not be able to do any kind of useful cost-time analysis of them.

Method of cost estimation

The tool does not do any kind of cost estimation.

Generation of financial indicators

Net present value (NPV)
Payback period
Internal rate of return (IRR)
Benefits cost ratio
Other

Ability to Include Environmental Costs

User friendliness and flexibility

The technique can be applied to a wide range of projects and processes. The software package includes a detailed discussion of the diagnostic techniques that users can study.

User-support

The tool was not designed to be a commercial product.

Limitations

The literature available expounds more on CTM as an analytical technique but does not really discuss the manner of use of the system (or related requirements and features). There is no discussion of anything other than usual costs. However, the system can be effectively applied to analyze hidden costs. The accuracy of the results will only be as good as the estimates and data supplied by users (on variables such as cycle time, wait time, costs, etc.). The package does not discuss the methodology that users may use to obtain this data (especially, estimating time). The system mainly provides users with a graphical "snapshot" of their process/project and a means of comparison with various alternatives. Cost details cannot be seen from the graphic. The tool will probably mask a great deal of information while dealing with large projects or those with long time horizons. These projects will need to be effectively broken down to a suitable level of detail.

Basis for evaluation: Manual on the CTM technique obtained through Westinghouse in February 1995.

Contact information: Available, though not a commercial product. Stephen Green ,Westinghouse Electric Corporation Environmental Affairs 11 Stanwix Street, Pittsburgh, PA 15222
412-642-2455/642-4027
Ginny Clanchard ,Westinghouse Electric Corporation Productivity and Quality Center PO Box 160 Oakdale, PA15230-0160 ,412-778-5169

EasyABC Plus

Prepared by: ABC Technologies. Inc.

Prepared for: Public purchase

Application: EasyABC Plus is an activity-based costing (ABC) system.

Elements of control:

Cost control
Resource control
Estimating control
Schedule control
Scope control
Risk control

Development date and updates: Released in 1990. Updates are - once a year. The current version is 2.05, which was released in 1995. The software is continually updated with new tools added to make ABC and ABM functions such as data collection easier.

Public availability: Available

Purpose and current use: The system is targeted to estimators, architects, and engineers. There are over 2000 copies of the software installed. A bulk of the clients have been from large companies, but small-medium sized businesses have also started purchasing the system.

Cost information: The approximate cost of software and support for EasyABC Plus is $5,500. The cost of OROS starts from - $15,000.

System requirements: This software is available in Windows, OS/2, and Macintosh versions.

Minimum system requirements for EasyABC Plus:

· DOS 3.l/MS Windows 3.x /OS/2 3.0/Apple Macintosh Finder 6.l/System 6.02 'DOS/Windows: 640K memory. OS/2: 4 MB. Macintosh: 1 MB
· DOS/Windows: 1 MB free disk space for installation. OS/2: 2 MB for installation Macintosh: 1MB for installation. For all versions, 6-8 MB recommended for operation 'DOS/Windows: EGA/VGA display. OS/2: OS/2 compatible display. Macintosh: Mac compatible display
· Mouse (Windows 3.x/OS/2/Macintosh compatible)

Minimum system requirements for OROS

· Windows 3.1
· 4 MB memory (8 MB recommended)
· Hard drive with 12 MB available
· VGA monitor
· Windows 3.x compatible mouse or pointing device

Software Summary

Size and complexity of projects: The system can include an unlimited number of activities. Multiple projects can be analyzed using the system.

Other compatible systems offered: OROS ABCPlus, a network version of the software, with additional features included.

Non quantifiable information: Model and period notes can be entered. This software enables users to develop and modify ABC models and generate information necessary to examine processes and to make informed decisions to improve them. The system provides feedback about improvement efforts and their effect on the bottom line. It helps users to recognize sustaining activities that give them a competitive advantage, while weeding out high-cost activities. The program comes with a proprietary database. The system helps users to organize information by dividing it into three categories that are included as three basic modules:

· Resource module: This reflects the general ledger items, such as accounting, engineering, sales, general administration, information systems, etc.
· Activity module: Activities can be grouped by department, where actual activities are identified through interviews.
· Cost object module: Cost objects are products or services that use up activities. These can be measured through drivers.

Although these modules are built individually, users use the same process to build each one. Information in the three modules can be linked in two different ways. The first is the **Output Measure Method.** This determines the unit cost of an activity, then adds the activity to a product to determine the product cost. The second method is the **CAM-I or the Cost Assignment Method.** This involves distributing overhead expenses to accounts, then to activities, and finally, to the cost objects. The software automatically makes connections and assignments based on information fed in by users. Features of EasyABC Plus include the following :

· Resources, activities, products, and services can be tagged with unique attributes, that can be used for analysis and comparisons
· Unit costs calculation for each resource, activity, product, service, and attribute
· Generation of performance measures that are cost per unit for an activity
· Adding, deleting, or renaming drivers (output measures)
· Incremental cost allocation, allowing calculation of a single account, a particular module, or the entire ABC model
· Any number of levels of cost assignments between resources, activities, or products
· Data storage for more than one time period and period-by-period report generation
· Displays of actual and budgeted costs (performance against plan comparisons)
· File security

A network version of the software, OROS ABCPlus, is now available, which has additional features such as multi-user capability, order-of-magnitude costing, and free-form reporting.

Life-cycle stages covered

Raw material acquisition
Manufacturing stage
Use/reuse/maintenance
Recycle/waste management

The system is not designed for life-cycle cost analysis. The life-cycle stages included in the analysis would be based on the company, or the type of project being analyzed. Hierarchies can be defined in the various program modules. However, these hierarchies are not tailored to define project breakdowns.

Type of costs considered

Conventional
Potentially hidden
Contingent
External

Activity based costing systems encourage significant breakdown of work activities and the proper allocation of costs, automatically making a number of potentially hidden costs more visible. External or contingent costs could be included/identified by specifying cost categories or additional activities. However, these are not included as default categories in the program.

Method of cost estimation

The systems are not designed for cost estimation.

Generation of financial indicators

Net present value (NPV)
Payback period
Internal rate of return (IRR)
Benefits cost ratio
Other

Ability to Include Environmental Costs

User friendliness and flexibility

Both systems have import/export capabilities, however, they are more simplified in OROS. EasyABC Plus supports import and export of ASCII and DIF files. It provides seamless interface for data interchange between leading process modeling tools. EasyABC Plus includes 13 standard reports that allow flexibility in the type and amount of information to be reported. OROS includes OROS Reports, a visual design report writing tool that supports ODBC and allows users to create custom and standard reports from the databases. Using the *OROS Links Engine* (provides open database connections), users can connect information already in their business networks (spreadsheets, databases, host applications) to the activity database. Concurrent access to a single activity database can be accomplished through OROS.

The user-interface and supporting documentation adhere to the CAM-1 standards of terminology. On-line help is available. The products run on all general purpose platforms and also meet all Windows and Macintosh user interface standards.

User-support

Support is available by fax and through a toll-free number, and a two-day training course is offered. Demonstration disks are available. However, these are not interactive, working demos. The company includes documentation, tutorials, and toll-free customer support. Training workshops are also available. At the conclusion of these workshops, users are given a free ABC kit that includes 60-day trial use of the software, a discount of $795 from the price of the software, and a tutorial. Also included are a video entitled *Understanding Activity-Based Management,* a checklist for implementing an ABM project, a seminar workbook, a disk with a working ABM model, and an ABC demo disk. The workshops are organized on different themes, ranging from general to specific. Consulting support is not offered directly through the company. However, ABC Technologies has partnered with several consulting firms who are trained in using their software. These consultants can offer basic training on the development of an ABC system, and methods of data collection.

Limitations

There is no specific focus on including environmental costs and cost categories in sample projects and in default formats. For example, less-tangible or contingent costs are not included as default categories in the program. However, the program can serve to automatically make some hidden environmental costs more visible.

Basis for evaluation: Based on information package received from ABC Technologies in June, 1995.

Contact information: ABC Technologies, Inc. 5075 S. W. Griffith Drive, Suite 200 Beaverton, OR 97005
(503) 626-4895 (503) 671-1808 (503) 626-4003 (FAX) Rick Musser
Information on factory support or training: 800-882-3141

Parade

Prepared by: Primavera Systems, Inc.

Prepared for: Public purchase

Application: Cost control and performance measurement. It is not designed to include all cost categories.

Elements of control:

Cost control
Resource control
Estimating control
Schedule control
Scope control
Risk control

Development date and updates: First release was in 1987. Updates - every 16 months. The current release of Parade is version 4.0, released in April, 1995.

Public availability: Available

Purpose and current use: Cross industry applications. Approximate number of users (not sites) of Parade are 7000.

Cost information: $ 3,000 for single user application

System requirements: The system requirements are: IBM PS/2 or PC compatibles
· 640 K hard drive MS DOS 3.1
· EGA/VGA monitor

Parade may also be used on a network (Novell, LAN manager, Banyan, Pathworks, LANtastic).

Software Summary

Size and complexity of projects: Various sizes and complexities

Other compatible systems offered: Parade can utilize data from Primavera's P3, Project Planner for Windows, Finest Hour, and Suretrak Project Manager.

Non quantifiable information: Provides logs for recording project notes.

The Parade software is for measuring performance and allows users to track progress against the plan using earned value analysis. It can be used for planning or for period-by-period control. Parade uses WBS (20 levels possible) and OBS (20 levels possible) structures and supports earned value analysis for the project as a whole or for any work element or group. The package has 60-year accounting calendars with 12 or 13 month years. Some features of Parade are as follows :exception reporting based on variances from plan at any level; reports and graphics including a responsibility assignment matrix (department/person intersections with work areas), performance bar charts and curves, variance trends in cost and schedule, cost performance and budget reports (summarized at any level of detail); variance detection with defined thresholds for favorable and unfavorable deviations from plan; and compatible with C/SCS criteria: DOD 5000.1 and 5000.2, DOE 4700.1, NASA 533P;

Parade can consolidate cost data from accounting, schedule data from a scheduler (P3, Finest Hour, and SureTrak Project Scheduler users have special integration features), and also draw estimates from spreadsheet or database files. Thus users have a choice of either building a project in Parade or importing data from elsewhere. Parade can automatically import resource allocations, WBS codes, rate escalations, and resource distribution curves from P3.

Life-cycle stages covered

Raw material acquisition
Manufacturing stage
Use/reuse/maintenance
Recycle/waste management

The system is not designed to consider more than one life-cycle stage, and it will typically be used to consider the manufacturing stage of the life cycle. Parade uses WBS (20 levels possible) and DBS (20 levels possible) structures, and this is one way that life-cycle information can be entered into the system.

Type of costs considered

Conventional
Potentially hidden
Contingent
External

Cost accounts in Parade can contain detailed, period-by-period task and resource plans. Parade tracks direct labor, material, equipment, travel, and other costs. The analysis data are presented in units, direct costs, or burdened costs. However, flexibility is afforded to users because it is possible to have unlimited resources per task and unlimited tasks and work packages per project. Additional cost categories cannot be created in Parade, and codes cannot be assigned to work packages.

Method of cost estimation

Parade is not designed for project cost estimation. It automatically calculates estimate to complete and latest revised estimate costs for on-going projects.

Generation of financial indicators

Net present value (NPV)
Payback period
Internal rate of return (IRR)
Benefits cost ratio
Other

Ability to Include Environmental Costs

User friendliness and flexibility

Parade can integrate cost and schedule data from a variety of other accounting, cost, and scheduling systems. In a multi-user environment. Parade permits simultaneous updating, reporting, and plotting.

User-support

Telephone hotline support and bulletin board. Working demonstration disks are available for Parade, limited by the maximum number of activities (60).

Limitations

The system does not include all the cost categories. User-defined cost categories cannot be added into the system. Parade is not designed for Life-cycle cost analysis.

Basis for evaluation: Based on information received from Primavera Systems in February 1995.

Contact Information: Primavera Systems Inc.
Two Gala Plaza Gala Cynwyd, PA 19004
610-667-8600 610-667-7894 (FAX)
Nicole Stephana 610-660-5830

ABBREVIATED PROFILES

FOREFRONT

FOREFRONT is a cost-tracking/accounting system targeted to the construction industry and project managers. It is not a cost estimating system. It allows users to enter in one set of dimensions and generate all the quantities, material prices, and labor hours for each activity defined. Crews can be set up with different labor and equipment rates. Bids can be readily and rapidly revised. The software produces cost bid summaries and subcontractor summary reports.

FOREFRONT allows users to define 60 different cost types that can be used for exception reporting and viewing. This will enable users to define and track a full range of environmental costs using the system. However, all cost categories are not included among the default data. Maintenance and repair costs are included in the demo data provided. Jobs can be broken down into phases (up to 12 characters). Any number of multiple projects and activities per project can be tracked.

The software can be customized to meet the needs of individual users. Users can obtain the source codes to alter the software. FOREFRONT has multiple user capability. Current developments include developing the database (Proisam from McDonald Douglas Info. Systems) and incorporating a project management capability. FOREFRONT has full ODBC capabilities, which means that data can be exported and imported freely. In addition, there is a special import module that ties in with Hard Dollar EOS, and with the Historical

Cost Analysis System (HCAS). FOREFRONT can also be integrated with an inventory system.

FOREFRONT was introduced in 1984. Most of the users are in the construction and environmental industry (for remediation work), there being about 500 clients in all. There are approximately three revisions per year. Version 5.0 was released on February 1,1995. Demonstration disks are included, but these are not interactive. However, live software demonstrations are offered by the vendors. User support is offered in the form of toll-free support, annual maintenance, and on-site training. No consulting services are offered.

FOREFRONT consists of 22 modules (General ledger, accounts payable/receivable...), each of which cost approximately $1,350.

Minimum Hardware: IBM PC with hard disk or equivalent; will run on UNIX, ZENIX and AIX. The system can be used on Netware.

Dorter & Chancy Business Systems Inc. 3200 NE 125th St. Seattle, WA 98125 206-364-1400 800-875-1400 206-367-9613 (FAX) Brad Mathews

Basis for information: Based on excerpt from Watson (1995) and on information obtained from supplier in July, 1995.

P/X

P/X is a Windows-based comprehensive high-end project management system. It can handle an unlimited number of inter-related projects and fully integrates schedule and cost data via resource-activity relationships. While P/X is not an "estimating" system, it will track costs against the baseline estimate by utilizing user-defined cost categories. The program contains a proprietary database. P/X is an open client/server application that sits atop popular commercial SQL databases (Oracle, Sybase, SQL Server, DB2/2 and others). It also contains earned value techniques to support the requirements for C/SCSC cost control.

P/X can be tailored to the unique needs of users with the optional P/X Customization toolkit. Users can add additional cost categories. The system can handle unlimited projects and activities per project. An unlimited number of WBS structures per project can be defined, with each having a maximum of 20 levels. All these features can facilitate a comprehensive inclusion of environmental cost data. The system has full import/export capabilities, using user- defined formats. Sample projects are included. There are no sample environmental projects.

The software was initially released in June, 1992, with updates released twice a year (on an average). It is targeted to Fortune 500 Corporations and government clients. It has been used by DOE contractors including Westinghouse, Allied Signal, and others. The system has been used for environmental remediation projects as well. User manuals and tutorials are provided along with the system. Consulting and product support are offered.

Multi-user versions are sold, with a starting price of $40,000 for a five concurrent user license. Per user price drops significantly with volume.

Minimum Hardware:

- IBM 486 PC or compatible 16 MB RAM
- 15 MB hard disk space
- VGA and mouse
- Windows 3.1
- ODBC driver for selected SQL DBMS

Supported networks (optional) include Novell, IBM LAN Server, Banyan Vines, and MS Lan Manager.

Project Software & Development, Inc. 20 University Road Cambridge, MA 02138 (617)661-1444 Grace Ferry Jerry Schinski 201-909-3765 201-587-1336 (FAX)

Basis for information: Based on excerpt from Watson (1995) and on information obtained from supplier in July, 1995.

CHAPTER 5. RISK ASSESSMENT SYSTEMS

This chapter provides descriptions on five software systems that deal with risk assessment. These programs are :

1. MOUSE (Modular Oriented Uncertainty SystEm) *Project Risk/Contingency*

2. Independent Cost Estimating Contingency Analyzer (ICECAN)

3. Monte Carlo

5. Environmental Risk Management Information System (ERMIS)

GENERIC RISK ASSESSMENT

MOUSE (Modular Oriented Uncertainty SystEm)

Prepared by: Risk Reduction Engineering Laboratory, U.S. EPA
Prepared for: Use in Federal facilities and public purchase.

Application: A computerized uncertainty analysis system using a special purpose simulation language.

Elements of control: Cost control - Resource control - Estimating control - Schedule control - Scope control - Risk control - The system only passively controls risk (although it can assist in dynamic control).

Development date and updates: Latest version (Stage III) brought out in 1990

Public availability: Available

Purpose and current use: The uses of MOUSE in the EPA include establishing regulations for hazardous wastes in landfills and estimating pollution control costs.

Cost information: The operational manual and the MOUSE computer system on disk are available from the NTIS for a charge of \$80.

System requirements: MOUSE system utilities run on MS-DOS (version 2x or higher) based PCS, but the MOUSE programs themselves can run on any computer that has a FORTRAN compiler. If substantive simulation models are to be run on a PC, a math co-processor is recommended.

Software Summary

Size and complexity of projects: MOUSE can support the use of up to 200 variables.

Other compatible systems offered: None

Non quantifiable information: None

MOUSE is a Monte Carlo-based simulation system that is designed to deal with uncertainties in static mathematical models, such as a set of environmental engineering calculations. Unlike deterministic algebraic models, stochastic models contain input variables in the form of probability distributions that reflect the uncertainty about their values. There are different approaches to solving algebraic models that involve uncertainty. MOUSE uses a form of Monte Carlo simulation known as Model Sampling. This method is an attempt to address problems associated with traditional approaches to uncertainty. These include an inability to show a combined net effect of changes in all variables or the likelihood of various changes occurring together. Model Sampling involves the following:

1. It performs a random drawing of values for each input variable, from their probability distributions.
2. The models output is computed using each such set of values.
3. The process is repeated many times and the output from all the iterations are gathered in the form of a probability distribution. This transfers the uncertainties of the model inputs to the output, which can be studied and used in the decision making process.

After the correct input of every equation, AutoMOUSE gives users the option of entering more equations or proceeding towards analysis. When equation entry is complete, users have to specify the nature of distribution and parameters for all stochastic variables, with the help of a menu. For example, users can select a normal probability distribution and specify a certain mean and standard deviation. If it is not clear what probability distributions should be used, the Interactive Modeler for Probability (IMP), which is a MOUSE utility program, can perform the following functions:

1. Fit the data set to a classical distribution.
2. Integrate an empirical continuous distribution.
3. Analyze a bivariate distribution.
4. Analyze a dataset for autocorrelation.

MOUSE possesses all the features for substantive uncertainty analysis, such as built-in probability distributions, plotting and graphing capabilities, sensitivity analysis, and interest functions for cost analysis.

Life cycle stages

Raw material acquisition

Manufacturing stage
Use/reuse/maintenance
Recycle/waste management

Not applicable.

Type of costs considered

Conventional
Potentially hidden
Contingent
External

Not applicable. There are no cost categories or cost summary reports generated. MOUSE can be used to analyze a set of equations related to any type of project or cost types.

Method of cost estimation

The system is not designed for cost estimation. However, it can be used to incorporate the risk element, thus improving the accuracy of various cost estimates, which can also include liability costs and less-tangible costs.

Generation of financial indicators

Net present value (NPV)
Payback period
Internal rate of return (IRR)
Benefits cost ratio
Other

Ability to Include Environmental Costs

User friendliness and flexibility

In its third and current stage of development, MOUSE can be used by individuals with little or no knowledge of computer languages or programming or simulation. Yet it can be equally beneficial for experienced users. Unlike earlier models of MOUSE, the current version does not require users to write a FORTRAN program or perform other cumbersome and error-prone tasks. Using a program called AutoMOUSE, the current version actually writes all of the MOUSE program. Users only have to input the equations of the model and answer queries regarding the output desired. They also specify the probability distributions for uncertain (stochastic) variables in the model and arguments for these distributions. AutoMOUSE operates in two modes: nonexpert and expert mode.

MOUSE has many advantages over other major approaches to conducting Monte Carlo simulation analysis on algebraic models. The other approaches use general purpose computer languages, such as FORTRAN; general purpose simulation languages; and spreadsheet Add-on programs.

AutoMOUSE makes preliminary and detailed checks for errors during equation input and editing. The scans check for syntactical and also logical errors (using a trace). The model equations can also be entered via a file prepared with any ASCII text editor. In the expert mode, users can make other choices such as specifying the number of iterations in the Monte Carlo simulation, performing sensitivity analyses for all or any of the independent variables and changing the random number seed.

User-support

Although it is available commercially, the system was not designed for that purpose. Hence, little user support (as offered by private vendors) is provided. Queries can be directed to the EPA Risk Reduction Engineering Laboratory.

Limitations

Each equation is limited to three lines. One of the disadvantages of MOUSE is that the programs have to undergo preliminary processing, such as compiling and linking, before they can be run. Advantages include its flexibility and ability to run complex standard simulation routines. AutoMOUSE is limited to a maximum of 200 variables, but it is stated that this does not pose a problem for most models.

Basis for evaluation: Based on information received from the EPA Risk Reduction Lab.

Contact information: Commercially available through NTIS.
EPA, Risk Reduction Engineering Laboratory 26 Martin Luther King
Cincinnati, OH 45268

Contact person: Albert Klee 513-569-7931
National Technical Information Service (NTIS)
703-487-4660 800-553-6847

Project Risk/Contingency Independent Cost Estimating Contingency Analyzer (ICECAN)

Prepared by: USDOE/IT Corporation

Prepared for: Public purchase

Application: Determining cost contingencies. Cost contingencies due to environment-related variables could also be analyzed.

Elements of control: Cost control - Resource control - Estimating control - Schedule control - Scope control - Risk control - The system only passively controls risk (although it can assist in dynamic control).

Development date and updates: -1983. The program is not periodically updated. It was updated to run using QBASIC and no longer runs on BASICA or GWBASIC.

Public availability: Available

Purpose and current use: A cost-estimating contingency model developed for the DOE. Typically used at the end of a remedial design phase. The system is being used in DOE and by various DOE contractors.

Cost information: The program is free.

System requirements: IBM PC running under DOS 2.0 with a minimum of 64K RAM and at least one disk drive.

Software Summary

Size and complexity of projects: Limited by the maximum number of variables (approx. 20) and distributions (4) that can be defined.

Other compatible systems offered: None (although IT Corporation has developed other software tools for DOE).

Non quantifiable information: None

ICEAN produces a frequency distribution of total costs based on several random cost variables. The cost variables can be fixed, normal, discrete, or step-regular.

Life-cycle stages covered

Raw material acquisition
Manufacturing stage
Use/reuse/maintenance
Recycle/waste management

Type of costs considered

Conventional
Potentially hidden
Contingent
External

The system has been developed for considering conventional costs alone, but users can analyze other costs by using the proper variables and relationships.

Method of cost estimation

The system uses the Monte Carlo method to compute random variables for the total cost function. The total cost function is normally simply the sum of the random cost variables. The mean of the sampled value of the cost function can be used as a reasonable cost estimate. However, the main purpose of the program is to estimate contingencies. Users can determine the probability of project overrun as a function of various contingency percentages. The model results are displayed in the form of a frequency distribution report and a contingency report.

Generation of financial indicators

Net present value (NPV)
Payback period
Internal rate of return (IRR)
Benefits cost ratio
Other

Ability to Include Environmental Costs

User friendliness and flexibility

Software options are accessed through a main menu. Options on this menu allow users to enter, edit, store, load, and delete files containing cost variable data. The software allows users to create one file for each estimate. Estimate processing time can range from a few minutes to over an hour depending on the number of random values (iterations) taken. Environmental costs can also be analyzed using the proper variables and relationships.

User-support

The user's manual describes the operation of the ICEAN program with regard to data entry, editing and filing, user options, and reports generated. Since ICEAN is not a commercial program, user support will be limited to responses to minor queries and information requests.

Limitations

Users do not have a wide choice of distributions by which to describe the cost variables. The ICEAN program is only as accurate as the description of variables provided by users. The distributions of critical cost variables must be carefully analyzed and require good estimating Judgment. While items such as spent dollars and firm commitments are fixed values, items such as direct labor and equipment may be prone to a high degree of risk.

Basis for evaluation: Based on profile prepared in WEC. Additional information was obtained through communications with IT Corp.

Contact information: Not a commercial package. Available to private industry through request from DOE (usually DOE contractors) W.E. Gallagher/Susan Watson IT Corp. (formerly PEI Associates Inc.) 4425 West Airport Freeway, Suite 350 lrving, TX 76062
214-570-2000 214-258-0399 (FAX)
Juan Castro
US Department of Energy 202-586-9706

Monte Carlo

Prepared by: Primavera Systems, Inc.

Prepared for: Public purchase

Application: Project risk analysis and simulation. It is not clear if risks due to environment-related factors can be considered, or if contingent environmental costs (e.g., liability costs) can be calculated.

Elements of control:

Cost control

Resource control
Estimating control
Schedule control
Scope control
Risk control

Development date and updates: The first release was in 1990. The current release is version 2.0, released in 1993. A Windows version of the program is expected to be released in the end of 1995/early 1996.

Public availability: Available

Purpose and current use; Cross industry. Monte Carlo has - 1000 users (not sites). **Cost information:** The system costs - $5,000.

System requirements: The minimum hardware requirements are an IBM PC (or compatible) with 80386 and 80387, 4 MB RAM (minimum), and DOS 3.1.

Software Summary

Size and complexity of projects: Various sizes and complexities

Other compatible systems offered: Primavera Project Planner (P3-Project Planner for Windows), Finest Hour, and SureTrak Project Manager.

Non quantifiable information: Import text files
The Monte Carlo software is for analyzing risk and mitigation measures. It uses the Monte Carlo simulation method to help quantify the effects of the many variables that can affect the outcomes of a project. It can be linked with Primavera Project Planner (P3-project

planner for Windows) project data to analyze expected dates, costs and critically develop contingency plans or make go or no-go decisions. Features of Monte Carlo include the following:

· calculation of variables such as expected cost, expected completion date, criticality indices and confidence intervals
· generation of reports and graphics for schedule, resource, and cost analysis.
· "quickrisk" feature allows assignment of global probability factors to project schedule and cost projections
· "cost time envelope" feature combines possible finish dates with corresponding costs to provide a series of time/cost end points for the project data transfer (expected dates and costs) back to Primavera Project Planner

The system also has many other features such as variable resource availability, user-defined correlations among cost elements and work efforts, probabilistic resource rates, and others. The reports and graphics for cost analysis include cost histograms and history graphs, cost probability curves, and cost/performance risk reports.

Life-cycle stages covered

Raw material acquisition
Manufacturing stage
Use/reuse/maintenance
Recycle/waste management

The system is not designed to consider more than one life-cycle stage, which would typically be the manufacturing stage of the life cycle.

Type of costs considered

Conventional
Potentially hidden
Contingent
External

It has been designed to consider only conventional costs. The accuracy of estimates if applied to other types of costs or in estimating liability costs is unclear.

Method of cost estimation

The system is not designed for cost estimation.

Generation of financial indicators

Net present value (NPV)
Payback period
Internal rate of return (IRR)
Benefits cost ratio
Other

Ability to Include Environmental Costs

User friendliness and flexibility

Unlike some other risk assessment tools on the market, Monte Carlo is geared specifically towards uncertainties associated with project management. It has many additional features such as stochastic analysis of critical activity paths, resource pools, unit cost tables, and probabilistic and conditional branching rules. The reports and graphics are also tailored for project managers.

User-support

Telephone hotline support and bulletin board service. A working demonstration disk of Monte Carlo is available.

Limitations

The system has many different features but is also more expensive than many of the basic risk assessment tools that are available. It does consider all cost categories.

Basis for evaluation: Based on information received from Primavera System.

Contact information: Primavera Systems Inc.
Two Bala Plaza Bala Cynwyd, PA 19004
610-667-8600 610-667-7894 (FAX)

Range Estimating (REP)

Prepared by: Decision Sciences Corp. (DSC)

Prepared for: Public purchase

Application: Risk analysis tool. The system can consider conventional, hidden, and contingent costs. However, contingent costs do not necessarily include environmental liability costs, and the accuracy of estimates in that regard has not been determined. These cost elements can include non-environmental factors as well. There is no specific guidance provided on how to include or handle environmental cost elements.

Elements of control:

Cost control
Resource control
Estimating control
Schedule control
Scope control
Risk control

Development date and updates: Introduced in 1972. The current release of REP is version 4.0, released in 1993.

Public availability: Available

Purpose and current use: The program was developed as part of a 7-year research effort, which began in 1968 and covered a broad spectrum of industries. REP has users both in the U.S. and abroad. It is used by private industry and government agencies. It has been used for a variety of problems including detailed cost estimating, negotiation, competitive bidding, project management, engineering design, new product evaluations, and research and development.

Cost information: The current REP for PC licensing fee is $1,715.

System requirements: REP is available for IBM and compatible PCs. Although the system has been used on a network, it is not designed to be network compatible, and cannot be accessed concurrently by multiple users.

Software Summary

Size and complexity of projects: No limit

Other compatible systems offered: None

Non quantifiable information: None

Range Estimating is a tool for looking at risk and sensitivity of cost estimates. The Monte Carlo Simulation method requires users to identify probability distribution functions for critical elements as well as parameters such as the mean and standard deviation. DSC believes this method can be very difficult to use effectively by inexperienced users. They believe that Range Estimating provides the solution by synergistically combining Monte Carlo simulation, Pareto's law, and DSC-developed heuristics to quantify and rank risks and opportunities for decision making. (Heuristics: Relating to or using a problem-solving technique in which the most appropriate solution of several found by alternative methods is selected at successive stages of the program for use in the next step of the program - *American Heritage Dictionary, Second College Edition*). Range Estimating only requires users to enter a range to identify and quantify risk. By ranking the input elements, it allows users to see why a decision's risk profile looks the way it does. REP can assist users in doing the following:

· analyzing contingency for detailed and conceptual estimates.
· developing bidding strategies for fixed price and unit price contracts
· performing project selection and ranking between alternatives
· performing strategic and long-range planning
· communicating risks to involved parties
· developing communications between departments and management concerning
project viability, estimate reliability, and project progress

REP can produce five management reports, a modeling report and four technical reports. The reports are simple and useful for managers (e.g., to estimate contingency, risks associated with each critical element, and the probability of having a cost overrun).

Life cycle stages

Raw material acquisition
Manufacturing stage
Use/reuse/maintenance
Recycle/waste management

The program is not designed to consider more than one life-cycle stage.

Type of costs considered

Conventional
Potentially hidden
Contingent
External

Any type of positive or negative flow can be analyzed using REP, as long as it can be conceptualized and a probability distribution can be drawn for it. The user's manual discusses the division of all costs into three different types: known, known-unknown, and unknown-unknown. The known elements affect the bottom-line in ways that can be measured. The known-unknown comprise those elements that are known to affect the bottom-line in ways that cannot be measured exactly. These would relate to hidden costs. The unknown-unknown cannot be analyzed, as there is no known methodology to handle them. Although the user's manual discusses all these cost types, there is no specific reference to environmental costs, and the importance of including them into analyses. For example, REP analyzes contingency for detailed and conceptual estimates. These are not necessarily contingent environmental costs, although environmental variables can be factored as critical elements into the analysis. Also, hidden cost elements would include environmental costs only at the user's discretion.

Method of cost estimation

The system is not designed for cost estimation. It is designed to improve on cost estimates by quantifying risk.

Generation of financial indicators

Net present value (NPV)
Payback period
Internal rate of return (IRR)
Benefits cost ratio
Other

Ability to Include Environmental Costs

User friendliness and flexibility

The system does not accept inputs from other software systems, but the manufacturers feel that this increases the productivity of the system by avoiding iatrogenic (induced by the assessor) risk. This happens because software systems often move too much, and the wrong data in the transfer process. This does not involve much extra effort on the part of users because most analyses do not have more than 10 to 20 critical elements. They also believe that the Monte Carlo method for risk analysis, especially when used on spreadsheets by nonstatisticians, can at times grossly understate risk. This happens when users identify too many variables in the spreadsheet. REP, on the other hand, uses Pareto's law to highlight elements that are incorrectly defined as critical by users.

User-support

The price of the program includes one copy of the software, user's guide, 90 days telephone support, and 2 years of program updates free of charge. Decision Sciences also offers a seminar that can be arranged at clients' facility at a fee of $1,400. No demonstration disks are offered, to protect the trade secret algorithms contained in REP.

Limitations

The program does not run on Windows but can run on any IBM-compatible computer without the need for a math co- processor or extended memory. DSC does not license its trade-secret heuristics to software developers for incorporation into their programs. Although REP can consider conventional, hidden, and contingent costs, these will not necessarily include environmental factors. DSC provides no specific references or guidance to users on how to include or deal with environmental variables and costs.

Basis for evaluation: Based on information received from Decision Sciences.

Contact information: Decision Sciences Corporation
Box 28848 St. Louis, MO 63123
314-739-2662 314-536-1001 (PAX)

ABBREVIATED PROFILES

Environmental Risk Management information System (ERMIS)

Woodward Clyde has developed a risk assessment tool for the

management of environmental financial risks. This method was initially developed in 1970, and has since been continually refined. ERMIS was developed in conjunction with a specific project. The software and services for liability costing are available to Woodward Clyde clients.

ERMIS uses a decision analytic approach to quantify environmental impacts, rather than environmental outputs alone. The purpose of ERMIS is to:

· Identify and describe high priority problems
· Quantify and describe, in financial terms, individual and aggregate environmental risks
· Summarize risk data by site, medium, and type of problem

Environmental risk quantification is typically established through a team comprised of a decision analyst, an environmental engineer, and a lawyer. The decision analyst structures the problem set and uses techniques including event trees, probability distributions, level of confidence calculations, etc. The environmental engineer estimates the costs and time frames of environmental damage and remediation alternatives. The liability lawyer estimates the costs and probabilities of compensatory damages, fines, and penalties. The process relics on available environmental data, and substitutes subjective expert opinions for hard environmental data, when that data is unavailable.

The environmental data and the probabilistic estimates collected through this process are fed into ERMIS, a relational database incorporating risk calculating algorithms and allowing flexibility in sorting and reporting. ERMIS calculates the probability of occurrence and ranks problems. Finally, estimates of the present cost of correction and the future liability from uncorrected problems are made. The present cost of correction and future liability are compared using varying timeframes and at varying levels of confidence. This helps in devising strategies and options for risk reduction. It is important that information derived be unbiased, as the initial data gathered is primarily subjective in nature.

Woodward Clyde feels that the expertise employed in liability assessment often more crucial that the software. Site specificity can make it difficult to make generic expert systems for environmental liability costing. This is because conditions are often unique and future outcomes cannot be determined from precedents. The cost of the service will be over $25,000 with typical costs being $40,000 - 50,000 (they can be much higher). The service has primarily been used by large companies.

James. H. Schaarsmith (Managing Director) Woodward Clyde Risk Management 500 12th Street, Suite 100 Oakland, CA 94607-4014 510-893-3600 510-874-3268 (FAX)

CHAPTER 6. REMEDIATION SYSTEMS

COST ESTIMATION AND SCHEDULING SOFTWARE

HAZRISK (database and models)

Prepared by: Independent Project Analysis, Inc. (IPA, Inc.)

Prepared for: A group of private companies and the U.S. DOE.

Application: A Parametric cost estimating tool to provide order-of-rnagnitude cost estimates for Environmental Restoration (ER) projects.

Elements of control:

Cost control
Resource control
Estimating control
Schedule control
Scope control
Risk control

Development date and updates: Research study began in 1988. Efforts are being made to expand the program to cover more projects.

Public availability: Available to the sponsors of the HAZRISK research program

Purpose and current use: The development of HAZRISK is funded by private chemical and oil companies and the US Army Corps of Engineers. The purpose is to provide industry and the government with improved methods of evaluating the cost and schedule risks (order-of-magnitude) of hazardous waste projects for both the Remedial Investigation/Feasibility Study (RI/FS) and Remedial Design/Remedial Action (RD/RA) stages of Environmental Restoration (ER) projects.

Cost information: A software package was developed for the private-sector sponsors and includes the models and database. The license fee for this program and joining the HAZRISK research consortium is $35,000.

Contents and system requirements: Models and proprietary database. IBM PC (or compatible) with Windows 3.1. The software has a worksheet to guide the user and provides results as probability distributions.

Software Summary

Size and complexity of projects: Primarily designed for larger projects.

Other compatible systems offered System available in different modules.

Nonquantifiable information Information not available.

HAZRISK gives users an understanding of the factors driving the cost growth and schedule slippage in ER projects. The database contains completed assessment and cleanup projects. The following type of information is contained in the database:

· general site information
· type of contaminant
· estimated and actual schedule

· cleanup technologies evaluated and selected
· project management information
· project funding information
· actual costs
· estimated costs
· regulatory and external events

Two types of models were developed and are used in combination to estimate project cost and duration and uncertainties surrounding the cost/schedule estimates:

direct cost and schedule estimating models-use Parametric techniques for estimation. · cost contingency allowance-predicts contingency allowance required at alternative levels of risk.

Costs and schedules are systematically linked with project characteristics such as the nature of the site, contaminants), technologies, regulations, etc. The links are forged using Parametric statistical techniques such as correlation and simple and multiple regression.

Life-cycle stages covered

Raw materials acquisition
Manufacturing
Use/reuse/maintenance
Recycle/waste management

Type of costs considered

Conventional
Potentially hidden
Contingent
External

Method of cost estimation

The database includes historical costs and schedules for completed remedial actions. Data sources included EPA, DOE, private-sector sponsors, engineering firms, and the Army Corps of Engineers.

The software has a worksheet to guide users. The required inputs are site characteristics, site cleanup history, cleanup technology, and nature and type of contamination. Parametric statistical techniques are used to establish links between these factors and cost and schedule.

Generation of financial indicators

Net present value (NPV)
Payback period
Internal rate of return (IRR)
Benefits cost ratio
Others

Ability to Include Environmental Costs

User-friendliness and flexibility

HAZRISK is useful when there are uncertainties in project and technology definition. It gives good early estimates for cost and schedules, using only a few data inputs. The purpose is to complement and not to replace design-based approaches. The advantages of the statistical models are that they do not require detailed data; they allow project risk to be treated and measured explicitly.

User-support

IPA has produced two studies that are available to the HAZRISK sponsors: The HAZRISK Assessment Study and The HAZRISK Cleanup Study. These describe the methodology and approach used in developing the HAZRISK models. The program sponsors in 1991, and an independent DOE peer-review panel (reviewed it from the point of view of applicability to DOE) validated the accuracy of the models.

Limitations

The assessment cost and schedule data are taken only from the RI/FS phase of an environmental remediation project. Cost and schedule numbers reflect the costs of remedial construction and treating waste. Long-term recurring expenses (site maintenance, monitoring) and estimates of hidden or less-tangible costs are not included. HAZRISK is only available to the research sponsors.

Basis for evaluation: The information provided here is based on a profile of the tool presented in the WEC (1992). Subsequent reviews were done by incorporating updated product information through telephone communications with IPA.

Contact information: Independent Project Analysis, Inc. (IPA, Inc.)
11 ISO Sunset Hills Road, Suite 300 Boston, VA 22090.
Dean Findley/Jenny Painter
703-810-1266 703-709-0777 703-709-0838 (FAX)

Remedial Action Cost Engineering and Requirements System (RACER)/ENVEST

Prepared by: Delta Research Corporation

Prepared for: U.S. Air Force, Directorate of Construction Cost Management (HQ AFCESA/CESM)

Application: Engineering-based environmental cost estimating system. Estimates costs for studies, remedial design, remedial action, and related site-work for environmental restoration projects. The target users are governmental and other remedial action estimators. RACER estimates capital and 0&M costs, contingencies, overhead and profit, and general indirect costs.

Elements of control:

Cost control
Resource control
Estimating control
Schedule control
Scope control
Risk control

Development date and updates: More remedial action models are being added to the program. Version 3.1 was released in July, 1995, and includes ordnance remediation, drilling, field sampling/mobile laboratory, preliminary assessment/site inspection, thermal oxidation, low level radioactive soil remediation, etc. (96 total).

Public availability: Available

Purpose and current use: RACER is in the third year of a continuing development effort. The software was developed for the U.S. Air Force Civil Engineering Support Agency (HQAFCESA/CESM) for use in environmental remediation projects beginning during the Preliminary Assessment/Site Inspection (PA/SI) phase and continuing through the RI/FS, design, and remediation.

RACER and its commercial version (ENVEST) have over 900 trained/registered users in government agencies (including the EPA), universities, and in private industry. Parties purchasing RACER as government contractors need to be working on contracts that require remedial cost estimating.

Cost information: $5,500 (11 disks). The approximate cost for a government contractor is $1300. The price includes software, documentation, system updates for one year, and help-line support.

System requirements: IBM PC 386 with Windows (version 3.0) 45 MB hard disk 4 MB RAM

Software Summary

Size and complexity of projects: Designed for medium-large sized projects.

Other compatible systems offered : The program is modular in design (and includes different models).

Nonquantifiable information : Information not available.

RACER was developed with the following objectives:

· budget preparation and cost comparison for treatment alternatives during the entire project execution.
· cost estimation and management
· collection and distribution of historical cost data and technical information

The RACER WBS is divided into five primary elements: PA/SI, RI/FS. RD, RA, and 0&M.

Life-cycle stages covered

Raw materials acquisition
Manufacturing
Use/reuse/maintenance
Recycle/waste management

The system is not meant for estimating costs of the entire life-cycle. The objective is to estimate remediation costs.

Type of costs considered

Conventional
Potentially hidden
Contingent
External

RACER estimates capital and operating and maintenance costs for remedial action treatment technologies. The cost categories that can be specified include equipment costs, material, and labor. It also estimates costs of studies, remedial design, and site-work and utilities. Contingency costs are estimated, but these do not include environmental liability costs. Users could detail additional costs by creating their own estimate (defining WBS levels). However, additional cost categories cannot be created, and any such cost and quantity estimates would have to be developed by users themselves. Lump-sum estimates of these costs can be included.

Method of cost estimation

RACER is a Parametric cost estimating system, which requires users to enter a minimal amount of information to generate estimates. The estimates reflect the engineering design of the process in question, rather than relying on historical cost data alone. The required parameters that users need to supply are project specific. The secondary parameters determined by the model, can be overridden by the estimator. RACER will supply default quantities based on similar project case histories and engineering assumptions. The predefined engineering relationships link the primary parameters with detailed engineering quantities. Finally, the quantities are priced using the cost databases. Assembly default values (quantity, unit costs, safety level) can also be changed by users. Cost data for RACER is derived primarily from the Army Corps of Engineers Unit Price Book, which is supplemented from other similar databases. In the last step, RACER provides estimates of Sampling and Analysis, RA professional labor, contractor general conditions, contractor overhead and profit, and Remedial Design. Multiple design options can be estimated to find the most cost effective solution.

RACER contains two study models (RCRA and CERCLA), a remedial design model, two project control models, fifty- five remedial action models, and thirty-six site work/utilities models. RACER cost estimates are location-specific and include contractor general conditions, overhead, profit, and escalation. RACER contains models to estimate low-level radioactive waste cleanup costs, but not for estimating mixed waste cleanups.

Generation of financial indicators

Net present value (NPV)
Payback period
Internal rate of return (IRR)
Benefits cost ratio
Others

RACER version 3.2 will include present worth calculations.

Ability to Include Environmental Costs

User-friendliness and flexibility

RACER allows users to modify any parameter or quantity within the system, including the default quantities and unit costs. System assumptions can also be analyzed and modified. Based on the information available, RACER can be used to prepare preliminary or final estimates. RACER estimates are more accurate than order-of-magnitude estimates, because they are based on engineering design and current cost data. The RACER cost models were derived using information from previous projects, government laboratories, construction management agencies, vendors, contractors, and engineering analysis. RACER system models use the Interagency Cost Estimating Group (ICEG) Code of Accounts.

RACER has interface options with other software packages, special reporting capabilities. The WBS can also be customized (to a limited extent) by users. Delta Research also provides options for customizing RACER, including adding technology models, expanding current models, adding system features, different model defaults, and links to software. In the next phase. Delta plans to enhance the user-defined model and templates to make the system more flexible, and to add present worth analysis capabilities.

User-support

There is an on-line help option that features graphical displays of (most) process designs. System training is offered at the training center in Florida and sometimes, at off-site locations. The class size is preferred to be between 5 and 15 individuals. In the commercial version of RACER (ENVEST), limited time program evaluations (15-20 days) are available.

Limitations

The RACER models and cost databases are limited to estimating conventional and operating and maintenance (hidden) costs. Other hidden, contingent, and less tangible costs are not estimated, but can be entered by users as a percentage of the total costs. Users could include their own estimates of these costs through the user-defined model. However, additional cost categories cannot be created. That is, users cannot have an additional category that details less-tangible or regulatory costs associated with various treatment technologies. RACER does not allow users to make changes in existing cost models (i.e., add their own algorithms).

Basis for evaluation: The profile presented here is based on product information provided by Delta Research Corporation in April, 1995.

Contact information: The commercial version of RACER is called ENVEST and is marketed by the Delta Technologies Group. Delta Research Corporation 1501 Merchants Way Niceville, Florida 32578

Engineering questions: Michael Reid

Software: Bob Smith

Special requirements: Scoff Henson

Cost estimating support: Michael Edgar

Customized RACER: Dan Murphy
904-897-5380

Remedial Action Cut Estimating System) (RACES)

Prepared by: International Technology Corporation (IT Corp.)

Prepared for: U.S. Environmental Protection Agency National Risk Management Research Laboratory (previously Risk Reduction Engineering Laboratory)

Application: Engineering-based, order-of-magnitude, cost estimating system for hazardous waste remediation.

Elements of control:

Cost control
Resource control
Estimating control
Schedule control
Scope control
Risk control

Development/publication date and updates: A "beta-version" of the RACES software was produced. This has a fully-functional user-interface and a control technology estimation module for final covers.

Public availability: Beta version available

Purpose and current use: Engineering-based cost estimating tool for hazardous waste remediation. The tool would have allowed users to build on engineering design and parameters (which other tools such as CORA did not do), the system was to be built in the context of the Army Corps of Engineers WBS and scheme to make RACES a fronted to MCASES. The preliminary estimate would be built in RACES, and the details would be built in MCASES (as remediation was not the main objective of MCASES). The development of RACES stopped before a complete version was built.

Cot information: Not applicable

Contents and system requirements: Several cost estimation models. System requirements: IBM PC/XT or PS/2 · I MB of hard disk storage 640K RAM MS/PC DOS 2.0 or higher

Software Summary

Size and complexity of projects: N/A. The development of the software system was not completed.

Other compatible systems offered N/A. The development of the software system was not completed.

Nonquantifiable information N/A. The development of the software system was not completed.

RACES was designed to include technologies/activities (called system modules) that can be used in Superfund site remediation. These were to include control technologies, treatment technologies, and general site work and site improvement technologies. Each technology/activity was to have a cost-estimating model. The system was intended as an interactive system for potential uses such as for developing scaping or budget estimates for planning purposes, comparing costs of alternative rertlediation technologies, calculating site-specific cleanup costs, checking cost estimates prepared by others, and conducting present-worth analyses.

Operation of RACES was designed to include three steps. First, general information about the site is entered, followed by module selection, and finally data entry, The first two steps are repeated as often as necessary. Each module would contain information on technologies (definition), module operation/input definition, output report, and bibliography.

A prototype of the system and one cost-estimating module have been developed, which includes details about ten technologies.

Life-cycle stages covered

Raw materials acquisition
Manufacturing
Use/reuse/maintenance
Recycle/waste management

The maintenance database would consider use of materials. In general, the system estimates costs of equipment and technologies for remedial/clean-up activities.

Type of costs considered

Conventional
Potentially hidden
Contingent
External

The system only considers equipment costs and hidden costs associated with maintenance.

Method of cost estimation

The system uses a line-item approach for costing and relics on unit cost, cost estimating relationship (CER), and operations and maintenance item databases. RACES would use a database, and users could tailoring input data to the specific technical requirements of the specific site. Default values could be used to get "ballpark" estimates. Use of RACES would be most effective in the later stages of a project, such as the design stage, when detailed data would be available, *he unit cost data would be from the RS Means company, but RACES was being modified to be compatible with the MCACES/UPS cost system. RACES' treatment technology costs are derived from CER curves.

Generation of financial indicators

Net present value (NPV)
Payback period
Internal rate of return (IRR)
Benefits cost ratio
Others

Ability to Include Environmental Costs

User-friendliness and flexibility

The tool was not fully developed, which would make it difficult to judge these attributes.

User-support

The National Risk Management Research Laboratory also oversees the "Superfund Innovative Technology Evaluation (SITE) Program." The SITE program conducts full scale demonstrations of innovative remediation technologies to collect cost and engineering data. They would have been in a good position to provide current and historical information on technology costs in RACES models.

Limitations

The level of detail required for estimates makes it more suitable for later stages of the design process. The development of RACES has completely stopped, and it is unlikely that the work will be resumed.

Basis for evaluation: The information provided here is based on a profile of the tool prepared in WEC (1992). Subsequent reviews were done by incorporating updated product information from reports and through telephone communications with the National Risk Management Research Laboratory April, 1995.

Contact information: The development of RACES is no longer an active prefect because of funding constraints, the beta version of the software is available for review from the project manager. Information about RACES:
Gordon M. Evans, Economist U.S. EPA National Risk Management Research Laboratory 26 W. Martin Luther King Dr. Cincinnati, OH 45268
513-569-7684

RI/FS: Planning and Risk Analysis
RAAS, MEPAS and ReOpt

Prepared by: Battelle, Pacific Northwest Laboratories

Prepared for: Department of Energy (DOE)

Application: To assist professionals at each stage in the Remedial Investigation/Feasibility Study (RI/FS) process.

Elements of control:

Cost control
Resource control
Estimating control
Schedule control
Scope control
Risk control

Development date and updates: Latest versions (version I of RAAS) available in April 1995.

Public availability: Available

Purpose and current use: ReOpt and MEPAS have been available as stand-alone applications and are already in use in the DOE, EPA, and industry. RAAS has not yet been released.

Cost information: The total cost of ReOpt (for Federal contracts) obtained through the DOE'S Energy Science and Technology Software Center (ESTC) is approximately $600. Commercial versions of ReOpt may be obtained through Battelle. Cost estimates for commercial versions are not available.

System requirements: System requirements for RAAS:

· IBM-compatibte (386) computer
· 8 Me RAM
· 20 MB available hard disk storage space
· 31/2 inch high-density disk drive
· Windows 3.1
· VGA monitor, and mouse.

ReOpt is available for both Macintosh and Microsoft windows platforms.

MEPAS can run on IBM-compatible systems with 640 K RAM, 20 MB hard disk, and a 132 line printer.

Software Summary

Size and complexity of projects: Information not available

Other compatible systems offered : ReOpt and MEPAS

Nonquantifiable information: Decisions are made using both quantitative and qualitative data. For example, users can identify and document facts, assumptions, and operating constraints. Information on any separate note-writing capability is not available.

Remedial Action Assessment System (RAAS), MEPAS (Multimedia Environmental Pollutant Assessment System) and Remedial Options (ReOpt) are designed to assist professionals at each stage in the Remedial Investigation/Feasibility Study (RI/FS) process. The development of RAAS began in 1985 and is expected to be completed in April 1995. The RAAS software system will include the following:

1. ReOpt, an electronic encyclopedia that provides detailed information about contaminants, mediums, and environmental remediation technologies.
2. MEPAS, a physics-based risk computational model.
3. A site description module that helps users quantify site data and define cleanup strategies, constraints, and assumptions.
4. A report writing feature that allows users to document information.
5. Simulations of over 100 state-of -the-art technologies.
6. Information on hundreds of contaminants.

RAAS is designed to help users understand what, how, why, and when environmental remediation alternative can effectively solve site-specific problems. RAAS assists in users in the following:

1. Analysis of preliminary data (prior to availability of actual site data) and identification of other data required. Also helps users identify and document facts, assumptions, and operating constraints.
2. Estimation of baseline risk through user-specified data in MEPAS portion 2. Detailed user-specified information is used for creating holistic cleanup alternatives (though ReOpt).
3. Identification of feasible remediation options and comparison of performance of preliminary alternatives can be made using criteria such as cost, residual contaminant concentrations and volume, toxicity and mobility reduction. Defining remedial options consists of linking feasible technologies to find combinations that will meet objectives.
4. Evaluation of complete remediation alternatives

Users supply RAAS with site-specific data, operating assumptions, and cleanup objectives. These are further refined through a series of questions presented by RAAS. A short list of clean-up alternatives is given to users (based on one or more environmental remediation technologies), which users can analyze in depth. RAAS allows users to analyze various alternative scenarios and helps them decide which technology (or combination of technologies) will be most effective in site cleanup.

ReOpt, a component of RAAS, is like an electronic database of information on remedial action technologies with auxiliary data on contaminants and regulations. It assists users in identifying potentially appropriate alternatives. ReOpt includes references to previous applications of remediation technologies and has as many as 500 bibliographic references for follow-up research.

MEPAS (Multimedia Environmental Pollutant Assessment System) is a human health risk model. MEPAS is a menu- run program and an enhancement of the Remedial Action Priority System (RAPS). It can be used at both active and inactive chemical and radioactive mixed-waste disposal sites. Within the RAAS methodology, MEPAS is used to estimate baseline and residual risk to potential on-site or off-site receptors associated with a hazardous and radioactively contaminated site. As a stand-alone application, it can estimate total population exposures associated with a contaminated site. MEPAS computes risk using empirical, analytical, and semi-analytical mathematical algorithms and methods specified in EPA risk assessment guidance.

Life-cycle stages covered

Raw material acquisition
Manufacturing stage
Use/reuse/maintenance
Recycle/waste management

RAAS looks at remediation effects on-site and after the contaminants are removed off-site (ex situ technology module). It does not consider how the contaminants were generated in the first place.

Type of costs considered

Conventional
Potentially hidden
Contingent
External

Costs are calculated based on algorithms associated with each remedial technology. RAAS does not use cost alone to evaluate alternatives. Thus, data on risk and effectiveness, for instance, are not converted into costs. The system mainly considers direct costs of particular remediation alternatives.

Method of cost estimation

Remedial alternative cost is estimated by combining RAAS methodology estimated parameters such as treated volumes, treatment times, and other appropriate factors with corresponding cost factors. Such cost and technology data are included as part of the individual technology modules in RAAS to support these alternative comparisons.

Generation of financial indicators

Net present value (NPV)
Payback period
Internal rate of return (IRR)
Benefits cost ratio
Other

Ability to Include Environmental Costs

User friendlincss and flexibility

RAAS has a graphical user interface that makes information easier to understand and use. It may be possible for a company to use MEPAS and ReOpt to estimate risks and liability costs of on-going projects, using hypothetical scenarios. ReOpt, for instance, contains data on regulations and technology requirements for remediation. However, these tools were not developed with that objective in mind.

User support

Though an annual maintenance contract, ReOpt is updated and redistributed every six months to reflect regulatory and technological changes.

Limitations

RAAS is not a tool to help users avoid generation of wastes or in waste management. It is designed primarily for problem solving. In the end the decisions between alternatives are not based on one number, rather a list of parameters that users can inspect to make their decisions. It is limited to remediation.

Basis for evaluation: Based on telephone communication with Janet Bryant (Battelle) and information package received in January 1995.

Contact information: Commercially available Battelle, Pacific Northwest laboratories Battelle Boulevard PO. Box 999 Richland.WA 99352
Janet L. Bryant Senior Research Editor RAAS Technology Transfer and User Training
509-375-3765 509-375-6417 (FAX)

COST TRACKING

Removal Cost Management System (RCMS) Version 4.2

Prepared by: Roy F. Weston Inc.

Prepared for: USEPA Emergency Response Division

Application: Cost accounting tool for Superfund sites. The system is designed to record conventional and hidden cost information for projects at Superfund sites. Contingent and less-tangible costs cannot be estimated. The system does not facilitate project scheduling nor does it allow project definition using a WBS. However, site costs can be estimated using historical and rate data, Archived cost information can be used to compare projected site costs with existing site costs. Archived information is not factored in only for on-going projections. The system functions more as a passive cost control system, through recording and monitoring of real time costs.

Elements of control:

Cost control
Resource control
Estimating control
Schedule control
Scope control
Risk control

Development date and updates: Developed 1986. Updated periodically, latest version is 4.2.

Public availability: Available

Purpose and current use: Developed by Roy F. Weston Inc., for the USEPA, as a cost management tool for Federal On-Scene Coordinators (OSCs). It is designed for government and contractor sites.

Cost information: Available free of charge

System requirements: Runs on a standard IBM PC and requires *040* K RAM with 560 K free. A rates disk (personnel and equipment cost information) for a site is required to run cost tracking.

Software Summary

Size and complexity of projects: The system could possibly support any project size. Incorporation of project complexities could be limited. The system is specifically designed to handle projects at Superfund sites.

Other compatible systems offered: The system has four modules: cost tracking, cost projection, invoice certification, and site summary, the cost tracking module can be used alone. The other modules require input from the cost tracking module to run.

Nonquantifiable information: There is a general comments submenu item in the 1900-55 menu item in cost tracking that can be used for writing memos pertaining to the site. Comments on personnel, equipment, and other field items can also be entered.
The system is used to track costs on a real-time basis while on-site and minimize unnecessary expenditures. Users track costs by entering known data inputs. RCMS comprises four modules:

1. Cost tracking: an accounting system to track personnel, equipment, and other direct costs.
2. Cost projection: an estimating system to project future costs based on information entered in cost tracking.
3. Invoice certification: for verification of contractor invoices.
4. Site summafy: for preparing contractor and USEPA/TAT reports using archived information. Users begin by adding a contractor or a government site, along with corresponding rate and related information for personnel, equipment, and other field costs. Rate information will appear automatically, if present in the rates disk, or else users have to enter it themselves. The cost tracking module uses cost codes and task codes. Cost codes show the cost types for each activity performed on-site and only appear in the "other costs" area. Task codes are defined by operation.

The transfer option allows the day's transactions to be transferred to another computer. Daily cost summaries under the three predefined categories (personnel, equipment, other) can be generated. Items that are first purchased need to be entered in the inventory status and do not appear in the daily report unless they are used. There are predefined reports in cost tracking, including 1900-55 Form, contractor cost log (site costs summary), and preview reports. The site summary module allows users various options to select categories and items for reporting purposes. For example, personnel summaries can be generated by category, individual, by task, etc. The invoicing module is designed for certifying contractor invoices by using archived information.

Life-cycle stages covered

Raw material acquisition
Manufacturing stage
Use/reuse/maintenance
Recycle/waste management

The system is primarily designed to monitor remediation Jobs. Since costs are monitored on a real-time basis, use/reuse and maintenance are taken into account. For example, health and safety costs and materials and supplies are accounted for. Equipment costs are tracked in monthly cycles.

Type of costs considered

Conventional
Potentially hidden
Contingent
External

RCMS is designed primarily to include information on conventional costs including labor, material, and equipment. Since the system is designed for remediation work, these costs will include some upfront hidden costs such as site preparation, engineering, and procurement, etc. There are pre-existing task codes for activities such as sampling, disposal, health and safety, and others. The first-level task codes cannot be chosen by users: they can only further define the task codes through a second-level code. The cost codes have to be used to enter cost information for these different tasks. Costs information is entered using the unit cost method. If the correct units are chosen, it may be possible to enter information on some other hidden costs as well, there is no framework to enter information on all contingent or external costs, unless they can in some way be linked with a suitable cost driver. However, the complexity of these costs may make this difficult. The cost projection module allows users to enter a percentage figure for site and contractor contingency. These figures are input by users and may or may not include contingent environmental costs. The system does not assist in the choice of these figures. Contingent costs cannot be broken down. Additionally, users have to choose from a list of pre-defined units (they cannot choose any unit). In any case, separate categories cannot be defined for these costs. Fields cannot be changed, and costs have to be detailed in predefined formats.

Method of cost estimation

The RCMS cost projection module estimates future site costs for both initial and on-going site efforts by using contractor and government rate information and historical cost data from the cost tracking module. The initial projection will use only the rate information and is primarily intended for estimating costs of a new site start-up. Archived site cost information can be used for comparison purposes. Projected costs for on-going sites are calculated using rates and archived information. The projection separates the contractor from government expenses.

Generation of financial indicators

Net present value (NPV)
Payback period
Internal rate of return (IRR)
Benefits cost ratio
Other

Ability to Include Environmental Costs

User friendliness and flexibility

The system is menu driven and methods of data entry are similar from one menu to another. There is an on-line help function. Cost tracking personnel, equipment, and other field cost entries can be marked as disputed and further flagged in the site summary and invoice modules.

User support

There is an RCMS support hotline to assist with user queries and problems.

Limitations

Users need to be familiar with DOS commands and ceding used throughout the system. The system does not have the Windows structure and has no mouse support. Installation of RCMS can at times present problems. Initial site information (name, location, number, etc.) cannot be changed once it is entered. The system uses traditional methods of accounting and does not encourage users to break down activities to a significant level.

Users cannot add additional fields, and costs have to be detailed in predefined formats. Contingent and less-tangible costs cannot be estimated or detailed. A percentage figure for contingency can be factored in cost projections, and this may or may not include contingent environmental costs. The system is specifically designed for government/contractor operations at Superfund sites. It may not be suitable for use in unrelated projects or by private companies. RCMS functions more as a passive cost control device and does not have functions that facilitate cost minimization, that is, it does not function to evaluate different scenarios and choices through viewing the effects on total costs. It can only indirectly foster cost control by providing better cost records and by estimating site costs. It is not meant for life-cycle cost estimation.

Date for evaluation: Based on information from RCMS version *4.2* user's guide.

Contact information : Available to private and government agencies upon request to USEPA Emergency Response division.
USEPA Emergency Response Division Washington. DC 20460
Bob Cibulskis USEPA/ERT. M/S 101 2890 Woodbridge Avenue Edison, NJ 08837
908-321-6746
Mefissa Bryant, R.0.W Sciences (RCMS Support Line) 800-999-6990

CHAPTER 7. TRACKING SYSTEMS

There are two general areas of software reviewed in this chapter, namely facility management and environmental health and safety, and those dealing with waste management and tracking. In the first category, the systems covered are AWARE, DataPipe, Environme ntal Management System,Facility Tracking System, the FLOW GEMINI Information System, and TINIA. Under the waste management and tracking category of systems, eleven systems are reviewed (Cradle to Grave Version 2.0, Docuwaste and other CHEMTOX systems, ENFLEX Data, G2 Real Time Expert System, Gentrax, HAZE, HAZMIN, Material Inventory Report System, PlantWare, WASTE DB, and Waste Trail). An abbreviated profiles section is also included in this chapter.

FACILITY MANAGEMENT AND HEALTH AND SAFETY

AWARE

Prepared by: RECRA Environmental Inc. (Chemical and Environmental Analysis Services)

Prepared for: Public purchase

Application: Environmental management-tracking and reporting. The system can track all conventional and hidden costs associated with environmental management in a facility. However, additional fields can only be added through RECRA. There are no fields for estimating or including information on less tangible or contingent costs, the system has a separate modulo that can be used to detail strategies on waste minimization projects which can further be fucked and analyzed using the system. However, there is no framework for project planning or option generation in the system. There is no feature for maintaining historical cost records.

Elements of control:

Cost control
Resource control
Estimating control
Schedule control
Scope control
Risk control

Development date and updates: Developed in 1989. There is a general release every I to 1.5 years. Version 2.5 was released in November, 1994.

Public availability: Available

Purpose and current use: The system is designed for cross-industry applications because it can handle all hazardous and nonhazardous wastes. Currently, the primary users are manufacturing companies, the system currently has about six users. However, this does not include out-sourcing. With this option, RECRA uses the AWARE system on a network (with the appropriate security) to maintain and handle data from different clients internally.

Cost information: The base price for the multiple-user license is - $25,000. The maintenance fee of $1,250 includes free updates and support line. The price is an approximate figure based on conversation with RECRA representative in April 1995.

System requirements: AWARE can run on a 80386 IBM-compatible computer with DOS 5.0 (or greater) and with a minimum memory of 640 K. The software includes Harvard Presentation Graphics' R&R Report Writer runtime, and FoxPro runtime. It can be run on a PC or on a network.

Software Summary

Size and complexity of projects: Not applicable: the system is not primarily designed for project management. It is meant to manage and control areas of regular company activities, *he waste minimization projects that are entered basically detail strategies that can be monitored using the system. Currently, users may not be able to size the system as required through purchase, but they can through the out-sourcing option.

Other compatible systems offered: System is modular in format.

Nonquantifiable Information: The module for waste minimization allows unlimited free-form text.

AWARE is marketed for use in private industry and by government facilities for meeting their environmental objectives including waste minimization toxics use reduction, and pollution prevention. The system comprises a series of modules that allow users to document, track, and account for raw material consumption (including energy) and total waste generation as a function of products produced. AWARE interfaces a database of waste characteristics with users' existing raw material and production data (can share data with existing system). Since AWARE can calculate loadings of individual components, users can track and account for where every targeted chemical originates and exists in the operation. The AWARE modules are as follows:

· **waste directory:** profiles waste materials and documents waste management practices
· **waste tracking:** produces a daily generation and movement of all wastes, including constituents and costs

· **waste accounting:** summarizes waste generation and handling over a specified period
· **raw materials and product directory:** supplements existing manufacturing data systems and defines, standardizes, and tracks the raw material usage through production
· **waste minimization:** allows the documentation and tracking of all pollution prevention opportunity assessments, source reduction codes, and strategies.

Various reporting requirements of federal and State regulatory agencies can be met using the system. It includes waste minimization and pollution prevention reporting capabilities (under EPA guidelines), which allow users to relate chemical releases to production activity, raw material usage, and costs. The waste minimization module does not allow users to actually plan projects or interactively facilitate decision-making. It provides a framework to include information on these projects, which can further be analyzed as it progresses using the data gathered in the AWARE system. Each project can be assigned a separate space in the module. Trend charts and other graphics can be utilized to analyze changes in quantities and costs over time.

Life-cycle stages covered

Raw material acquisition
Manufacturing stage
Use/reuse/maintenance
Recycle/waste management

The life-cycle stages covered by the software may vary by module. Most of the modules cover the waste management stage of the life cycle. The raw materials/product directory tracks raw material usage through production. A product-tracking module tracks the movement of products in a facility. All the constituents that make up raw materials and wastes are fucked. This is a unique feature of AWARE.

A CIS system uses AWARE data to compute material mass balances. An important feature of AWARE is that chemical releases can be linked to raw material usage, costs, and production activity. That is, actual quantities of waste can be linked to quantities of raw materials used or products. This can be accomplished for all waste products, including scrap. However, no sensitivity analysis can be done to model changes in waste quantities and costs as a function of source reduction.

However, these life-cycle stages are not necessarily covered in the case of each individual chemical, material, or product. For example, if a chemical 6 not manufactured in the facility, only use and waste disposal stages (for that chemical) would be considered. Also, the products manufactured in the manufacturing stage are not covered in all their life-cycle stages either The life-cycle stages are not kept distinct within the system.

Type of costs considered

Conventional
Potentially hidden
Contingent
External

Users can account for the quantities and costs associated with managing all documented waste. The system also allows users to allocate costs back to the department or source responsible for waste generation. AWARE can represent costing trends for corporations, plants, departments, operations, waste materials, and chemical constituents. It can also manage and track production data including profit and revenue per unit (not calculated). All management costs including storage, labor, equipment, capital costs, transportation, and others are covered. AWARE considers only conventional costs and hidden costs associated with waste management. It does not attempt to put a dollar value on contingent or external costs. Additional fields to cover these costs can be included only by customizing the system through RECRA.

Thus, the full costs of using a material or product in a facility, including all but two of these cost categories can be estimated using AWARE. However, the costs that are entered cannot be collated to form a historical) cost library. This would be useful in estimating costs in projects that propose to use those raw materials or for estimating actual savings or waste minimization that could result through source reduction or process changes.

Method of cost estimation

The system is not designed for cost estimation.

Generation of financial indicators

Net present value (NPV)
Payback period
Internal rate of return (IRR)
Benefits cost ratio
Other

Ability to Include Environmental Costs

User friendliness and flexibility

AWARE generates reports to suit both the technical and nontechnical audience. It is designed so that it can be used with ease by an inexperienced computer user. Users can add to some information in the system such as the chemical database, but no user-defined fields can be added.

User-support

RECRA offers a number of services to AWARE users including waste characterization, in-plant data collection as an input to AWARE, systems integration services, pollution prevention consulting, and personalized training programs. Remote demonstrations of the software are provided by RECRA (with enabling software also provided).

Limitations

User-defined fields cannot be added in the system. Contingent and less tangible costs cannot be estimated or detailed. There is no feature for maintaining historical cost records, as related to different activities, quantities, and types of materials managed. The system cannot integrate with company accounting systems. This can result in these costs being managed separately, rather than being integrated with other company costs. The system has a module for waste minimization project monitoring. However, it cannot asset as an interactive decision-making or project planning tool. Although the system is primarily designed to facilitate compliance, there is an effort to include cost management as part of the system. If this area is further developed to provide information for capital budgeting (for instance using historical cost trends), pollution prevention and other environmental initiatives would be encouraged as a result of more complete cost information.

Basis for evaluation: Based on telephone communication with RECRA representative and information brochure received in February 1995. Additional information obtained from Rich (1995).

Contact information: Commercially available
RECRA Environmental Inc. Audobon Business Center 10 Hazelwood Dr. Amherst.NY 14228-2298
Lori A. Gasiecki (AWARE support specialist)

716-691-2600 800-52-RECRA

DataPipe

Prepared by: Knorr Associates

Prepared for: Public purchase

Application: Environmental health and safety (EHS) and environmental software. As yet, the systern has no cortlponents for project nnanagertlent. Some cost fields exist, but cost control and management is not a primary function of the software. There is no feature for maintaining historical cost records. Customization of the system can be accomplished only through the vendor.

Elements of conlrol:

Cost control
Resource control
Estimating control
Schedule control
Scope control
Risk control

Development date and updates: Introduced - 1969, periodic updates. There is no specific version number as different modules are updated, as required, at different time periods.

Public availability: Available

Purpose and current use: The software is designed to be suitable for a wide range of companies. The clientele for the product are primarily manufacturing companies and utilities.

Cost information: The single user price for the core module is $1,995 and additional modules range from 51,000 to 55,000 (some are more expensive). The system can be used in a multiuser environment, where each additional user (in increments of *5)* adds 10% to the price. Prices are approximate values based on telephone communication with Knorr representative in March 1995.

System requirements: PC-compatible computers using DOS and MS Windows. Minimum requirements: PC 386 4 MB of RAM · 10 MB free space on the hard disk (for initial data storages is recommended.

The requirements vary depending on how the system is used. The system can be used on a LAN or a WAN. Supported networks include Novell, Vines, 3Com, IBM Token Ring, and LAN Manager (any Windows-compatible LAN should work).

Software Summary

Size and complexity of projects: Not applicable; the system is not designed for project management. It is meant to manage and control aspects of regular company activities. The modular design will probably support operations of different sizes and requirements. Data from different modules can be used together. The system allows users any type of Customization for reporting requirements. Different methods are available for searching records, and databases are pre-indexed for speed and cross-referencing. However, there are no user-defined fields, which can limit incorporation of unique requirements and complexities.

Other compatible systems offered: System is offered in different modules that allow users to build a system that suits their requirements.

Nonquantifiable information: Not applicable; the system is not designed for project management. Most DataPipe forms have one or more 'notes' fields that can hold free form text. The current limit on one such field in a record is 4,000 characters.

DataPipe has been designed to collect, integrate, recall, report, and analyze information on health, safety, and environmental issues that face a company. It has environmental, safety and industrial health, and medical modules. The areas covered by all the modules include the following:

· solid waste, hazardous materials and waste, storm water and runoff, wastewater, groundwater, pretreatment, asbestos, and emissions management
· permit tracking, source inventory, MSDS and right to know management, health and safety management, and training
· TSCA, Clean Air Act, and other State and Federal regulations
· cost management
· reporting-including SARA, NPDES, and Form R

The environmental modules include those for chemical inventory and agent mass balance, MSDS, training, spill calculator, NPDES and emissions, waste manifests and PCBs, and waste treatment and minimization. The safety and industrial hygiene modules include those for agent mass balance, agent use, cause codes, incidents, MSDS ingredients, occupational injury and illness, radiation sources, work history, and training. The medical modules include (among others) the injury and illness module, worker's compensation module, blood and urinalysis module, audiometrics module, and physical exam and physical history modules.

The software includes forms where information can be entered, edited, or displayed. The specific fields and the type of information that users can enter (e.g., text, numeric, dates) will vary according to the form.

DataPipe can store and search free-form text and can store images. Thus items such as photographs, facility maps, copies of permits, and MSDS sheets can be stored as images along with text and numeric data.

DataPipe has three report writers, two of which are included in the standard system. The first allows users to quickly design reports by selecting fields to print their order, selection criteria, etc. The second system is optional and allows users to draw report layouts on a form, pasting database fields as they want them to appear. The third and most sophisticated report writer allows any combination of any portion of data from DataPipe. Users have access to the database programming language, which allows them to prepare custom reports. The system includes common reports associated with each type of information.

Life-cycle stages covered

Raw material acquisition
Manufacturing stage
Use/reuse/maintenance
Recycle/waste management

Life-cycle stages covered may vary by module. All the modules would not cover these life-cycle stages, nor would they do so for all materials and products. The life-cycle stages are not kept distinct within the system. The use/reuse stage would be considered only within the confines of a facility. Materials and wastes may be tracked; however, all costs associated with this may not be recorded (or estimated). The chemical inventory module, for instance, tracks where (facility, department, area of building), and in what quantities, an agent is used. The waste TSD module can identify the sources of waste production within a facility. The container mass balance module tracks the movement of containers to and from spaces and the addition and removal of agents between containers. However, these life-cycle stages are not necessarily covered in the case of each individual chemical, material, or product. For example, if a chemical is not manufactured in the facility, only use and waste disposal stages (for that chemical) would be considered. It may not be possible to link the different life-cycle stages. That is, a no quantitative relationship can be made between the quantities and types of material used or type of production process and wastes generated. Thus, users cannot simulate changes in waste quantities and costs as a result of process changes or source reduction.

Type of costs considered

Conventional
Potentially hidden
Contingent
External

All costs are not covered. For instance conventional costs will include items such as worker's compensation and costs of dine visits. Hidden costs will include items such as costs of waste treatment, storage, and disposal (TSD). All operating and labor costs associated with environmental and health and safety management are not recorded. Users cannot themselves add additional fields to include costs that are not covered in the modules. External or contingent costs that may be involved in the various life-cycle stages cannot be estimated or entered.

Thus, the full costs of using a material or product in a facility, including all these cost categories, cannot be estimated. Also, all conventional and hidden costs are not fucked and entered into the system. The costs that are entered cannot be collated to form a historical cost library. This feature could encourage companies to take some hidden costs into account in capital budgeting or project management. For example, the chemical inventory module could have an additional feature to record costs associated with using an agent in different areas (facility, department, area of building), quantities, and costs incurred as a result of safety management, storage, etc. This could be useful in a project that proposes to use that agent for estimating the actual costs involved.

Method of cost estimation

The system is not designed for cost estimation.

Generation of financial indicators

Net present value (NPV)
Payback period
Internal rate of return (IRR)
Benefits cost ratio
Other

There are no financial functions within DataPipe. However, users can set up a Dynamic Data Exchange (DDE) within Windows, to send data to other applications that perform these financial functions.

Ability to Include Environmental Costs

User friendliness and flexibility

DataPipe has a graphical user interface, and most operations for performing particular tasks have been standardized to be identical in all modules. The system has import and export capabilities that make it possible to connect it with other computers (e.g., a corporate mainframe or an analytical lab) for exchanging information. DataPipe supports about a dozen different common formats, including delimited ASCII spreadsheet and dBase. Data can also be imported from instruments that either have an RS-232 interface, or can write to a PC-compatible floppy disk. The system has record locking and other features for multi-user security. The security tables can limit the menus and menu items, including forms, that users can select.

Each DataPipe database has preset indexes for speed and cross-referencing. All DataPipe fields (except images) can be searched, whether or not they are indexed. This can be done by setting low and high limits on numeric, date and time fields, text fields for using words or parts of words, or words that sound like the one being searched.

Although this system is not designed for project management, it does help companies maintain better information and control over environmental management. Companies may avert some contingent costs and less-tangible costs by using software systems such as these, but the costs and benefits cannot be quantified.

User support

The basic single-user system includes personnel, security, two report writers, menu management, and other common parts. It includes one year of free support, including technical assistance, maintenance, and upgrades. The network system has an additional per-user license fee (cost depends on type of network-e.g., LAWWAN). Annual support costs 12% of the current system cost per year. Services for custom development and modification of modules and training are provided at different negotiated rates. DataPipe supplies demonstration disks that explain the capabilities of the system and allow users to look at sample screens and displays. On'site demonstrations of the system can also be arranged.

Limitations

The fields in the various DataPipe forms are predefined and cannot be changed by users, and no additional fields may be added (unless users specifically request the manufacturer to modify the forms). DataPipe does not have features or existing fields for estimating, or including detailed information on all cost categories. External or social impacts are not quantified or considered.

Modules in DataPipe will consider a variety of different costs. However, users cannot detail these costs. The system has no feature for maintaining historical cost records, as related to quantities and types of materials managed or operations related to environmental management. The system cannot integrate with company accounting systems, which can result in managing these environmental costs separately, rather than integrating them with other company costs. The system is not designed for, and has no modules for, project management. Like most other environmental management software tools, the primary objective is to facilitate compliance, not to manage and minimize costs. However, if this area were developed, pollution prevention and other environmental initiatives and projects would be encouraged as a result of better cost information.

Basis for evaluation: Based on telephone communication and information package from Knorr received March 1995. Additional information obtained from Rich (1995).

Contact information: Commercially available
Knorr Associates 10 Park Place. PO Box 218 Butler, NJ 07405-0218
Pete Singer (Senior Analyst)
201-492-8500 201-492-0453 (FAX)

Environmental Managmmnt System (EMS)

Prepared by: Chemical Safety

Prepared for: Public purchase

Application: Erlvironmental health and safety (EH&S) software. The system has some fields where cost information related to purchasing and waste management can be entered. These costs cannot be detailed as required. Additional fields can be added only through special Customization service requests to Chemical Safety. There is no feature for maintaining historical cost records. As yet, the system has no components for project management.

Elements of control:

Cost control
Resource control
Estimating control
Schedule control
Scope control
Risk control

Development date and updates: The system was created at a research facility in the University of California at Berkeley.

It was available for public use in 1996. The system is updated every quarter.

Public availability: Available

Purpose and current use: Chemical Safety has about 40 clients using the system. Although designed for cross- industry applications, EMS has primarily been purchased by research laboratories, universities, medical centers and pharmaceutical companies. EMS is being used in some government agencies as well.

Cost information: Each of the EMS modular packages addresses specific compliance areas listed below with their prices.
1.Chemical Inventory Management/MSDS Access: *$2,500*
2. Waste Management and Reporting: $3,800
3. Human Resources Management: $2,800
4. Material Safety Data Sheet (MSDS) Management: $1,500

The prices listed are for total packages, but users also have the option of purchasing individual modules in each of these packages in which case the price would be higher. These prices were effective from March 1994 and are subject to change without notice.

System requirements: The minimum hardware requirements for EMS are: 386PC DOS 5.0 ;640 K memory
· 100 MB hard disk (with at least *75* MB space available)
· 4 MB RAM
· LAN (for multi-user installations).

The system works on PC-based and Windows-based environments and interfaces with various desktop, workstation, and mainframe-based computer platforms. It can be used in a multiuser, network-based environment.

Software Summary

Size and complexity of projects: Not applicable

The system is not designed for project management. It is rneant to manage and control areas of regular company activities. Since the system is modular in design, users could size it according to their requirements.

Other compatible systems offered

EMS can be bought as a complete integrated unit or in modular format.

Nonquantifiable information: Not applicable; the system is not designed for project management. Every database record has a note field attached to it.

EMS is designed to manage environmental compliance and management requirements related to EH&S. It makes use of a database of information that is updated at least annually. The product is not designed to generate options and does not do any risk assessment or liability costing. It has a required module for facility management, wherein all company-specific information related to chemical inventories and regulations is managed. This module has additional options for safety equipment (location and maintenance), import and maintenance of *CAD* diagrams, and safety plans listings for the company. The various EMS modules address the following areas:

· hazardous materials and waste and wastewater management
· permit tracking, source inventory, MSDS and right to know management, health and safety and training management, property transfers and audits, mapping, and GIS
· cost management
· reporting requirements, including SARA reports and Form R

EMS has components for Federal, State, and local regulatory compliance and reporting requirements for EPA and OSHA regulations. The tracking component allows users to maintain complete records of ail materials by location and department/person accountability. Built-in chemical reference tables are designed to include regulatory, hazard, and reporting data for EPA designated hazardous and extremely hazardous chemicals. The system supports waste minimization and source reduction by reducing material purchasing and disposal costs through surplus management.

The Waste Management System tracks hazardous waste handling from pickup to packing and disposal. It also prepares manifests, tracks permits, and prints drum labels. Label and storage requirements are graphically displayed. The waste management system also has a module for contract/cost/invoice management for waste hauling and disposal, allowing users to maintain records on contracts, cost-effectiveness, and other miscellaneous information. Chemical safety offers additional hazardous waste modules including those for PCB tracking, container tracking, generating land ban forms, and hazardous waste handling/transport took-up tables.

The MSDS management system allows users to develop and maintain MSDS sheets using libraries of standard phrases and sections. The MSDS database cross-references with a chemical reference database that contains hazard and regulatory information on over 4,500 substances. An MSDS management version has just been released in the Windows format, which allows users to import and catalogue MSDS images in addition to MSDS in a database format. MSDS sheets may be obtained and converted from a variety of sources including CD-ROM and printed MSDS's scanned through OCR.

The human resources management component includes employee training, industrial hygiene monitoring, and incident/accident management and reporting, which includes generating the OSHA 200 log.

The other major modules include air emissions. Rid ad hoe management reports, and ICS inventory collection system. The air emissions module gives users the necessary information required for the Air Toxics Release Form R. ICS is a data collection tool for chemical inventory data input, which also contains a database of chemicals with their CAS numbers and tables with vendor names, units of measure, container types, product class, etc. It is offered for both DOS and Macintosh systems.

Life-cycle stages covered

Raw material acquisition
Manufacturing stage
Use/reuse/maintenance
Recycle/waste management

Life-cycle stages covered may vary by module. The Waste Management System covers the fourth stage as it tracks hazardous waste handling from pickup to packing and disposal. MSDS management and the chemical inventory help manage use and reuse of chemicals in the facility (although this information is not tracked). The human resources management component also includes activities undertaken for environmental health and safety. The system covers the manufacturing stage only to the extent that byproducts (emissions) as a result of the manufacturing process can be computed. The waste management portion of EMS is jinked with the chemical inventory. Once a chemical is declared as a waste, it is transferred to the waste management system.

However, these life-cycle stages are not necessarily covered in the case of each individual chemical, material, or product. For example, if a chemical is not manufactured in the facility, only use and waste disposal stages (for that chemical) would be considered. Also, the products manufactured in the manufacturing stage are not covered (if at all) in all their life-cycle stages either. The life-cycle stages are not kept distinct within the system.

Type of costs considered

Conventional

Potentially hidden
Contingent
External

The system considers hidden costs related to hazardous waste management. The contract/cost/invoicing module could include information on some conventional costs such as tabor and materials as well. It includes an area for cost tracking in inventory that covers conventional costs related to purchasing chemicals and materials. The preloaded databases in EMS contain some default cost information. The system includes a table that allows users to enter information on waste amounts, which are converted into pounds, and disposal costs are estimated. EMS does not have a framework to estimate or include liability or external costs. All other hidden and conventional costs associated with use and reuse of materials, environmental management (e.g., capital, tabor), and compliance (e.g., health and safety, administrative costs) are not included.

Thus, the full costs of using a material or product in a facility, including all these cost categories cannot be estimated. Also, aft conventional and hidden costs are not fucked and entered into the system. Users can customize the system only through service requests to Chemical Safety. The costs entered cannot be collated to form a historical cost library. This could be a useful feature for assisting in cost estimation for projects, because hidden cost estimates cannot be found in most (if any) commercial cost databases.

Method of cost estimation

The system does not do any cost estimation.

Generation of financial indicators

Net present value (NPV)

Payback period
Internal rate of return (IRR)
Benefits cost ratio
Other

Ability to Include Environmental Costs

User friendliness and flexibility

EMS has the capability to interface with existing inventory, purchasing, employee and other systems. Users can customize almost any facet of EMS, including reporting and adding new fields and features to suit their requirements. EMS has a limited amount of flexibility in incorporating user-defined fields (usually up to three). The system has on-line help. The modular format allows users to purchase only the kind of information they currently require with the option to make additions in the future.

User support

The chemical database is compiled from Federal and State databases and is updated regularly. Upon request, Chemical Safety also provides a list of clients for reference about the product. They provide options for system support, consulting, software/hardware leasing, and inventory analysis and data entry. The annual system support fee is provided at I *5%* of the system license. Chemical Safety also offers services for system Customization and project/implementation management.

Limitations

Although cost information can be included in certain sections, it is not given special attention. Rather, the focus is on management activities that would help prevent unnecessary expenses. However, there is no format to evaluate these initiatives. The operational costs associated with tracking, human resource management, etc., are not included in the system. EMS 6 restricted to hazardous material and waste management. It does not address management of all solid waste, stormwater, drinking water, and others.

The system considers only hidden costs of waste disposal with no options for additional fields to be added by users. These cost components cannot be detailed as required by users. Other conventional, contingent, or less-tangible costs cannot be estimated or detailed. Changes in the system can be made only through service requests to Chemical Safety that are charged on a time and expense basis. There is no feature for maintaining historical cost records, as related to different activities, quantities, and types of materials managed. The system is not designed for, and has no modules for, project management. Like most other waste management and tracking tools, the primary objective 6 to facilitate compliance, not to manage and minimize costs. However, if this area were developed, pollution prevention and other environmental initiatives would be encouraged as a result of better cost information.

Basis for evaluation: Based on telephone communication with Chemical Safety representative and information package received in February 1995. Additional information obtained from Rich (1995).

Contact Information: Commercially available
Chemical Safety 1301 S. 46th Street RFS 180-11 Richmond, CA 94804
Joanne S. Levy (Marketing Representative) 510-231-5641

Facility Tracking System (FacTS)

Prepared by: Quantum Compliance Systems Inc.

Prepared for: Public purchase

Application: Environmental health and safety (EHS) management. The system has some components that are used in project planning and monitoring. However, it does not address areas such as project cost estimation, scheduling, etc. FacTs gives users the flexibility to detail environmental costs as required. These costs can be used to generate summary reports or exported to packages such as Excel for analysis. The archiving feature allows users to retain historical cost information, although only along with all other information.

Elements of control:

Cost control
Resource control
Estimating control
Schedule control
Scope control
Risk control

Development date and updates: Company founded in 1986. Product development date not available. The next version, version 7, of FacTs will be released in fall. 1995. This will be a graphical user interface version (GUI).

Public availability: Available

Purpose and current use: Systems are in use in the aerospace, automotive, electronics, pharmaceutical chemical, and waste treatment industries. Quantum Compliance has sold approximately 100 product licenses (about 1,500 users).

Cost information: Information not available

System requirements: Supports multiple platforms (e.g., PC-DOS, UNIX, VAX).

The system requirements vary. The product can be used on a PC or on a network.

Software Summary

Size and complexity of projects: The system is not primarily designed for project management. It is meant to manage and control areas in regular company activities. The modules for project management can help monitor projects of various sizes but due to their limited nature may not support all project complexities.

Other compatible systems offered: FacTS has 30 modules, which can be purchased as required beyond the foundation Base system. FacTs is built around a relational database that allows all these modules to function in an integrated manner.

Nonquantifiable information: Information on any annotation capability not available.

Quantum offers FacTS as a modular, configurable EHS information management system. This has been designed to run on multiple platforms and support multi-facility operations. Quantum offers families of products for users. These include the Facility Tracking system for manufacturing and service organizations; the Waste Information Management System (WIMS) for waste treatment, storage, disposal, recycling and transportation facilities: the MSDS Generation and Delivery system: and Computer-Based Training. FacTS modules include those for air emissions management, chemical inventory tracking, illness and injury tracking, MSDS template design, sample tracking, SARA reporting, and waste tracking. Areas addressed by the various modules include the following:

· hazardous materials and waste, asbestos, and emissions management
· permit tracking, source inventory, MSDS and right to know management, health and safety and training management, risk analyse, property transfers, and audits for TSCA and Clean Air Act, and cost management

Other features of the system include the following:

. facility-specific data and corporate roll-up reporting across multiple facilities
· fourth generation language and relational database technology
· flexible data sharing with other databases including MRP, inventory and purchasing and accounting systems
· interface with bar-ceding, image scanning, and fax transmission systems
· customizable menus, on-line context specific help, and reference libraries for standardized terms, policies and procedures
· over 600 standard (including SARA reports. Form R, NPDES) and unlimited custom reports
· data entry in the form of verbatim text, scanned images, codes, or combination

The reference library covers information on SARA, RCRA, CERCLA, TSCA, OSHA, CAA. CWA. 67/548/EEC. State, local and user-defined topics. The system does not address solid waste, stormwater, groundwater, wastewater, drinking water or pretreatment management.

FacTs has three modules that are designed to support project management planning and monitoring activities. These are primarily used to monitor work progress and 'performance against plan" for project areas carried out by different contractors. These three modules are **"Standard Operating Procedures," "Work Order," and**

"Information Requests." The "Standard Operating Procedures" module tracks and manages tasks or projects and defines procedures to be followed for their safe completion. The "Work Order" module generates work orders defining tasks, priorities, due dates, and resources required. Although not as detailed as other project management tools offered in the market, this area can help environmental considerations more integrated with projects. However, it is typically used in environmental projects (remediation related) and may not be utilized in other projects that only have environmental components.

Life-cycle stages covered

Raw material acquisition
Manufacturing stage
Use/reuse/maintenance
Recycle/waste management

Life-cycle stages covered may vary by module. The waste management module (WIMS) covers cradle-to-grave tracking of wastes. WIMS covers waste treatment, storage, disposal, and recycling. The MSDS and chemical inventory tracking considers use and reuse of chemicals. The manufacturing stage is considered to the extent that emissions as a result of process activities can be calculated. However, the actual products of manufacturing are not tracked (unless the company manufactures hazardous chemicals, for example). However, these life-cycle stages are not necessarily covered in the case of each individual chemical, material, or product. Also, the products manufactured are not covered (if at all) in all their life-cycle stages either. The life-cycle stages are not kept distinct within the system.

Type of costs considered

Conventional

Potentially hidden
Contingent
External

The software contains different fields where users can detail the types of costs, dollar amounts, and units. All fields are user defined and users can add fields to detail costs as required. Thus, users could develop a system to track and record all costs. However, the system still does not assist in estimating external and contingent costs. There is no feature for creating historical cost libraries (that can further be used in cost estimation). However, all information from FacTs can be archived. Also, the report writer included in the system allows users to create custom reports using any combination of fields. Mathematical equations can be created to perform various types of cost analyses. The cost fields can also be sent to other packages, such as Excel. ft may be possible to create historical cost libraries in this manner. A package like Excel will allow users to perform different types of graphical and financial analyses of the costs as well. However, there is no methodology to assist in dynamic cost control in the system.

Method of cost estimation

The system is not designed for cost estimation.

Generation of financial indicators

Net present value (NPV)
Payback period
Internal rate of return (IRR)
Benefits cost ratio
Other

No financial indicators are generated within the package. However, cost data can be exported to spreadsheet programs such as Excel that do have financial functions.

Ability to Include Environmental Costs

User friendliness and flexibility

The modular structure allows users to move from simple to advanced applications, as required. The products offer flexibility through Customization options, such as custom exception reporting, user-defined default settings, and menus. The relational database technology and integration with other systems simplifies data entry and affords a wider perspective for users. The report writer allows customized reporting using any field combinations.

User support

Quantum offers remote demonstrations of their products. The products come with user manuals, on-line documentation, and the option for an annual maintenance contract, which provides software updates and telephone support. Other services provided include system integration and installation, custom software development, training, and environmental engineering consulting.

Limitations

The system has some modules for project management, but these modules address limited areas. Users would not be able to build and manage a complete project using these modules. The system as used contains qualitative and quantitative information that can be useful in project management, such as hidden costs or safety considerations for material selections. However, it doesn't have an interactive function that allows this information to be processed for use in project planning. If this area is developed, the tool could prove useful for incorporating environmental costs and considerations into projects and decision making. Like most other environmental management tools, the primary objective is to facilitate compliance, although the tool allows detailed cost tracking. Users can add cost fields, but the system as is does not consider a comprehensive list of costs. Less-tangible and contingent costs cannot be estimated.

Basis for evaluation: Gased on telephone communication with Quantum representative and information package.

Contact Information: Commercially available
Quantum Compliance Systems Inc. 4251 Plymouth road. Suite 1200 Ann Arbor,MI 48105 313-761-2175 313-761-3058

FLOW GEMINI Information System-EIS, OHIS

Prepared by: GRC International, Inc.

Prepared for: Public purchase

Application: Environmental, and occupational health and safety (EHS) management. The current system does not consider environmental cost information. However, GRC plans to expand the system to make this feature available. As yet, the system has no components for project management.

Elements of control:

Cost control
Resource control
Estimating control
Schedule control
Scope control
Risk control

Development date and updates: Introduced - 1981, Updates twice a year.

Public availability: Available

Purpose and current use: GRC has targeted the product to chemical and pharmaceutical companies in the past. They also have clients in industries such as aerospace, construction, electronics, environmental, and manufacturing. They are breaking down the modules further to allow the system to be suitable to an even wider range of companies. They have not targeted Federal facilities as yet. The system has some 50 clients.

Cost information: Prices are based on individual modules. A range of prices of all modules is $5000 to $15,000 (stand alone). Concurrent user pricing is available for networks, as wall as site licensing. The site license (unlimited users) typically ranges from $15,000 to $90,000 (depending on modules purchased).

System requirements: The system can be used by itself or on a network. Flow Gemini offers Client/Server versions for PC. LAN, UNIX, DEC, and IBM.

Software Summary

Size and complexity of projects: Not applicable: the system is not designed for project management. It is meant to manage and control areas in regular company activities. Current system is suited to large operations. GRC plans to scale down the system to suit smaller companies and operations as welt.

Other compatible systems offered: Modular format (can be purchased separately over the core modules or in an integrated form).

Nonquantifiable information: Not applicable; the system is not primarily designed for project management. Users can write operation/company specific notes (in MS Word).

GRC offers two basic systems to users: Occupational Health and Safety (OHIS) and an Environmental Information System (EIS). The EIS covers areas such as chemical tracking, SARA reporting, monitoring, and waste management. The areas covered by all the modules include the following:

· solid waste, hazardous materials and waste, storm water and runoff, wastewater, groundwater, and emissions management
· permit tracking, source inventory, MSDS and right to know management, health and safety management, and training
· Clean Air Act (CAA) management
· reporting-including SARA, NPDES, and Form R
· on-site waste minimization and treatment

EIS has modules/areas for SARA compliance, monitoring, permit tracking, and waste management. Chemical tracking functions include mass balance capabilities, material tracking (e.g., receipts, storage, transfers, process consumption, product output, releases, shipments, waste generation and disposal) for intra-facility locations, and relationships between manufacturers, suppliers, product codes, materials, and inventories. The SARA compliance function allows users to maintain Federal and State chemical list information, information management for emergencies, and reporting (Tier I and 11 and Form R reports in EPA formats). Information about samples (collection conditions, preparation, chain of custody, laboratory information) from any type of environmental media and analysis results can be logged, including a complete description of any type of release source. Information concerning any type of permit can be managed in the system including basic descriptive information and parameter limits (maximum, minimum, and average values can be logged according to permit, regulatory, and/or corporate specifications). Flow Gemini catalogs all wastestreams and includes information on amounts, profiles, on-site minimization, treatment, containers, storage, and shipping records. It generates the Uniform Hazardous Waste Manifests and manages information for annual waste generator reports.

The OHIS has a medical module, an industrial hygiene and safety module, and a hazard communications module. The medical module helps comply with OSHA requirements and maintains information on employee health, training, work history, and exposure. It automates equipment and medical supply inventories. The industrial hygiene and safety module is designed to determine compliance with OSHA, generate and track requests for MSDSs (supports ANSI Z400.l format), comply with community right-to-know requirements, document sampling results, track employee information, and audit laboratories. The hazard communications module generates and tracks requests for MSDSs, schedules employee training, and documents workplace inventories.

RAMS, the Remedial Action Management System, is an optional module or stand-alone system that automates the processes involved in monitoring chemically contaminated sites and RCRA facilities. A new MSDS system will be released in March 1995. This will support the ANSI Z400.l format for recording and printing, unlimited query capabilities, multilevel security, a graphical user interface, expanded data dictionary, flexible customizable templates, multiple notes features, flexibility in searching and reviewing MSDSs, editing capabilities including global updates, and FAX and e-mail support.

Life-cycle stages

Raw material acquisition
Manufacturing stage
Use/reuse/maintenance
Recycle/waste management

Hazardous wastes can be tracked (through a transaction-based management system) from the point of generation in the facility to receipt of a certificate of disposal from a TSDF. The system can track wastes back to a process or project. All the modules would not cover these life-cycle stages nor would they do so for all materials and products. For example, if a chemical is not manufactured in the facility, only use and waste disposal stages (for that chemical) would be considered. It may not be possible to link the different life-cycle stages. That is, a no quantitative relationship can be made between the quantities and types of material used or type of production process and wastes generated. Thus, users cannot simulate changes in waste quantities and costs as a result of process changes or source reduction. The different life-cycle stages cannot be kept distinct in the system.

Type of costs considered

Conventional
Potentially hidden
Contingent
External

The current system does not handle environmental costing. Only some conventional costs such as worker's compensation and employee assistant program referrals can be entered. The next release will make environmental costing part of the standard system. The company plans to have features that allow integration of the system with a company inventory/accounting system. Users will be able to detail costs.

Method of cost estimation

The system is not designed for cost estimation.

Generation of financial indicators

Net present value (NPV)
Payback period
Internal rate of return (IRR)
Benefits cost ratio
Other

Ability to Include Environmental Costs

User friendliness and flexibility

Flow Gemini is a client/server application that uses Oracle or other industry standard relational database management systems (RDBMS). Since the product has an open-relational database structure, it can export and import data to/from other relational database files, spreadsheets, and other files in ASCII format. It has a menu-driven/windows format. The system has a built-in image viewer. The default database currently contains information on areas such as materials tracking, mass balances, etc. They plan to include cost information in this as well. Flow Gemini comes with a suite of tools for customizing and reporting, which gives clients the ability to customize the system to incorporate their own terminology and business standards. The current MSDS module has query capabilities on any number of fields. Multilevel security can be provided. The system covers a very broad range of issues related to environmental management within a company. The RDBMS structure will allow companies to integrate other systems that may cover areas that Flow Gemini does not address.

User support

GRC customizes the database on specific requests from users. For example, the system database can be customized to add costing (not part of the standard system). The company provides customer support and training classes for all levels of users.

Limitations

The system does not currently handle cost information although GRC anticipates including this component in the next release. The present system only has fields for some conventional costs. It does not include a feature for maintaining historical cost records, as related to quantities and types of materials managed. The system is not designed for, and has no modules for, project management. Like most other environmental management software tools, the primary objective is to facilitate compliance, not to include cost elements and detailing. However, if this area were developed, pollution prevention and other environmental initiatives would be encouraged as a result of better cost information. The system is presently too expensive to be suitable for small to medium users, but the company plans to break down the modules further to make using the system feasible for small to medium companies.

Basis for evaluation: Based on telephone communication with GRC representative, and information package.

Contact information: Commercially available
GRC International, Inc. 1900 Gallows Road Vienna, VA 22182-3801
Karen Ward 800-934-5917 703-760-8747 (FAX)

TINIA

Prepared by: Alternative Systems Inc. (ASI)

Prepared for: Public purchase

Application: Environmental health and safety (EH&S) management. All areas of TINIA have a financial component that helps in tracking various costs related to EH&S. Users can add fields to include costs that may not be covered. There is no feature for maintaining historical cost records. The system has some limited components for project management. These are primarily meant for project monitoring rather than planning. The modeling and risk assessment functions in TINIA could be used in project planning and assessment applications.

Elements of control:

Cost control
Resource control
Estimating control
Schedule control
Scope control
Risk control

Development date and updates: Developed in 1986. Periodically updated. Version 3.3 was released in the first quarter of 1995. A graphical user interface (GUI) version is expected in 1996.

Public availability: Available

Purpose and current use: Applicable in a wide range of industries. Other information on specific industries and number of clients not disclosed.

Cost information: Information not available

System requirements: TINIA can be run on many different platforms and operating systems (the system is built on ORACLE).

Software Summary

Size and complexity of projects: The system is not primarily designed for project management. It is meant to manage and control areas of regular company activities. The system is probably more appropriate for larger operations. It can be used to monitor pollution prevention projects.

Other compatible systems offered: The system is offered in modular form. However, ASI advises against users buying individual modules (although it is sometimes sold that way).

Nonquantifiable information: Information on any annotation capability is not available.

TINIA is an Oracle-based integrated and flexible tool for environmental health and safety and information management. It focuses on the organizational issues that surround the management of hazardous waste and strategies to fulfill these objectives. TINIA was initially introduced in 1989 to help industry comply with Federal, State, and local regulations. It can be used on a LAN. TINIA has functions that address industry, government, and Federal facilities; thus, it can be used as a compliance tool and also to track facilities and waste sites while monitoring their compliance and remedial status. TINIA has eight major functions (modules), which are as follows:

1. Material Index: Covers chemical information, MSDS management, spill modeling, and reporting, etc.
2. Facility Management: Covers facility information and history including compliance history, complaints, regulatory fee, and fine tracking, etc.
3. Release Management: Covers biological/chemical and physical

releases into all environmental media (air, water, and land). This features air emission factors/calculations, sampling and analysis, leak detection, and repair, etc.

4. Material Management: Manages from purchase request to disposal including shelf life management, material waste mass balance (materials are balanced against the wastestreams and environmental releases they generate), container transfers, process safety, process costing/profitability, etc.

5. Waste Management: Covers waste cost tracking, pollution prevention, waste exchange, etc.

6. Human Resource: Covers workplace health and safety including employee information, education and health, workplace monitoring and controls, and risk assessment.

7. Emergency management: Covers emergency plans and response actions including material behavior and impact models, sensitive receptors, and decontamination procedures.

8. Commitment management: Includes functions for scheduling and communications.

All these functions are linked into a central relational database and common information is shared among all functional areas, thus reducing data entry, permitting global changes, and facilitating the sharing of traditionally segregated information.

TINIA can generate all U.S. Federal and most State, local, and regional reports.

The pollution prevention component in the waste management module allows users to set up strategies, goal, and baselines. The projects can be set up and exported to other software such as Primavera, for detailed cost estimating, scheduling, etc. Project progress against plan can be monitored within TINIA.

The scheduling function is not as sophisticated as other scheduling software available. Critical paths cannot be determined. It is used to track and meet day-to-day schedule requirements, including compliance report submissions, etc.

Life-cycle stages covered

Raw material acquisition
Manufacturing stage
Use/reuse/maintenance
Recycle/waste management

Life-cycle stages covered may vary by module. TINIA can track chemicals from cradle to grave. However, this means from the use/reuse stage (including maintenance) to the point of waste generation and ultimate disposal point. Wastestreams can be managed, taking into account treatment, storage, recycling, and destruction. For regulators, TINIA tracks facilities and waste sites through all phases of compliance. They may also track waste shipments and monitor environmental releases.

The process mass balance function can be used in conjunction with the modeling function to simulate changes in waste quantities and costs as a result of process changes or source reduction. Thus, a quantitative linkage can be developed between the quantities and types of material used or type of production process and waste generated, which can be useful in evaluating pollution prevention projects. The actual products of the manufacturing process are not tracked.

Type of costs considered

Conventional
Potentially hidden
Contingent
External

All areas of TINIA have a financial component, which helps in tracking capital costs, operating costs, efficiency, and return on investment (ROI) relating to EH&S. Thus the impact of EH&S on the overall profitability can be gauged. Costs can also be assigned to fugitive emissions. Process profitability is gauged on the basis of all conventional and hidden costs; by tracking inputs, outputs, and all equipment; and material and labor costs associated with a process. Also, each screen has a user-defined field function that allows users to add fields to cover additional costs or cost categories. TINIA does not include a means for estimating contingent and external costs, but it recognizes that proper management reduces potential liability and enhances company images. Additionally, the risk assessment function in the human resources module could be used to estimate contingent costs. However, users would need to develop their own methodology to accomplish this.

Method of cost estimation

TINIA is not designed for cost estimation.

Generation of financial indicators

Net present value (NPV)
Payback period
Internal rate of return (IRR)
Benefits cost ratio
Other

Ability to Include Environmental Costs

User friendliness and flexibility

ORACLE is platform-independent: therefore, TINIA can run on many different platforms and operating systems. TINIA can be coupled with several interface products, for example with imaging applications. Geographic Information Systems (GIS), Optical Character Recognition (OCR), and graphics. Modifications to TINIA are easily achieved by the open database architecture. Thus, organizations can make appropriate changes and additions to the system as they expand. TINIA is planning to convert to ORACLE Version 4.0, which will give users a complete graphic interface and mouse support.

Although the primary objective is to facilitate compliance, there is an effort to include cost management and minimization into the software features. Interface with company accounting systems is provided as an optional feature.

User support

ASI provides each client with a needs assessment service that helps guide them through all phases of implementation. They also offer training and consulting support services. Multilingual support is offered for international clientele. Remote demonstrations of TINIA are offered (required software is provided).

Limitations

TINIA does not have a methodology for material selection based on costs, quotes, or attributes. However, the modeling capability could be applied in conjunction with the process mass balance function to compare waste outputs to material use in hypothetical situations. The system considers various conventional and hidden costs with options for users to add additional fields. It provides no methodology for estimating contingent or less-tangible costs and has no feature for maintaining historical cost records, as related to different activities, quantities, and types of materials managed. The system is not designed for project management. It does have an area for recording and monitoring pollution prevention projects. This area should be further developed to use historical cost and other information for project planning, thus allowing users to use information that traditional cost-estimating tools typically leave out.

Basis for evaluation: Based on telephone communication with ASI representative and information package.

Contact information: Commercially available
Alternative Systems Inc. (ASI) 225 S. Cabrillo Highway Suite 124-C
Half Moon Bay, CA 94019
Jay D. Rades (Environmental Research Scientist) Michael Brinck (Senior Marketing Executive) 415-726-5700 415-726-7846 (FAX)

WASTE MANAGEMENT AND TRACKING

Cradle to Grave Version 2.0 (Terralink Product Series)

Prepared by: Terralink Software Systems Inc. (previously Brand Software)

Prepared for: Public purchase

Application: Hazardous waste tracking. As yet, the system has no components for project management. A limited number of environmental costs can be entered, though not detailed. There is no feature for maintaining historical cost records.

Elements of control:

Cost control
Resource control
Estimating control
Schedule control

Scope control
Risk control

Development date and updates: Introduced - 1993. Last update was in June, 1995, when version 2.0 was released. Updates are done once a year, on an average.

Public availability: Available

Purpose and current use: Terralink is considering targeting the product to large companies in the chemical industry. It can be used by generators, transporters, brokers, and TSDFs. Being a relatively new release, the system currently has a client list of under a dozen companies.

Cost information: The price of the basic program is $5,000. The average sale price for the small to medium companies has been observed as approximately $14,000 (single user). For full service transporters approx. $5,000 to 9,000, and for TSDFs as $9,000 to 12,000. Prices are only approximate values based on conversation with Terralink representative in March 1995.

System requirements: The system is IBM-PC based for single-user applications. Supported networks (LAN and WAN) include Novell, Windows for workgroups, and LANtastic.

Software Summary

Size and complexity of proiects: Not applicable; the system is not designed for project management. It is meant to manage and control areas of regular company activities. Available in different modules that could allow the system to be scalable and flexible enough to be suited to single- user small or large operations.

Other compatible systems offered: System is offered in different modules.

Nonquantifiable information: Not applicable; the system is not designed for project managernent. Note-writing capability in every section of the program.

Terralink's product series is a Windows-based, wide ranging modular tool for solid and hazardous waste tracking. The modules include a base module (Cradle to Grave) and subsequent tracking module (TerraTrac), profile module (TerraPro), communications module (TerraCom) and a manifesting and profile satellite. The areas covered by the various modules include the following:

· hazardous material and waste management
· cost management
· over 60 standard reports and customized report writing

Terralink's product series deals with profiling, manifesting, and tracking waste. Wastestream paths can be analyzed and details can be entered on waste profiles, chemical information, manifest numbers, shipment dates, facilities, exposure amounts, and final disposal methods. It includes reference libraries such as DOT hazardous material tables and EPA waste codes. It also includes a complete TSDF library. The system can track waste by process, department, or unit responsible for generation. It helps in compliance with CERCLA and RCRA regulations. Reports generated include Land Disposal Restriction Forms, transporter fee reports, and monthly and quarterly state reports.

TerraCom (through modem interface) facilitates information exchange between generators and transporters or between various offices. The manifesting satellite allows companies to print manifests in a remote location while keeping the information in a central office. The profile satellite can be used to gather profile information from remote locations and can be installed on a laptop computer.

Life-cycle stages covered

Raw material acquisition
Manufacturing stage
Use/reuse/maintenance
Recycle/waste management

The system currently tracks waste materials from the point at which they enter the system, to when they turn into waste and are finally disposed of at their final resting place. Although Terralink's product series can track waste by process, department, or unit responsible for generation, it does not detail these elements. There is no quantitative linkage between the quantities and types of material used or type of production process and waste generated. Thus, users cannot simulate changes in waste quantities and costs as a result of process changes or source reduction. There is no tracking of raw material and chemical use in production, or other activities. All materials and outputs are not tracked. Recycling is not taken into account.

Terralink is planning to release an inventory module, TerraTory, which will additionally include the use/reuse stage of the life-cycle. This module will track all incoming raw material throughout its life in a facility (including waste disposal and recycling), and each section will have a costing portion. This will also include a library for MSDSs.

Type of costs considered

Conventional
Potentially hidden
Contingent

External

The system has three costing fields in which users can enter hidden cost information associated with waste disposal, such as transporter and manifest fees. Thus, only a limited number of costs can be entered. Other conventional costs, such as labor, capital, and material costs that companies would need to incur during the course of environmental management activities, cannot be detailed. External or contingent costs that may be involved in the various life-cycle stages that hazardous materials would go through cannot be estimated or entered. The actual cost involved in the tracking activity is not included.

Thus, the full costs of using a material or product in a facility, including all these cost categories, cannot be estimated. Also, all conventional and hidden costs are not tracked and entered into the system. The costs that are entered cannot be collated to form a historical cost library. This could be a useful feature that could encourage companies to take some hidden costs into account in capital budgeting or project management. For example, if a particular project involved purchasing and using a hazardous material, they would be able to estimate some hidden costs that could be expected for material tracking and regulatory compliance, etc.

Method of cost estimation

The system is not designed for cost estimation.

Generation of financial indicators

Net present value (NPV)
Payback period
Internal rate of return (IRR)
Benefits cost ratio
Other

Ability to Include Environmental Costs

User friendliness and flexibility

The program is interactive in nature. It is written in FoxPro (now a Microsoft product) and therefore can access data from other Windows products. Each section has ten extra information fields that users can fill according to their requirements, providing some flexibility to users.

Terralink's product series comes with a built-in reference library that includes the DOT Hazardous Materials Table, EPA waste codes, and the Emergency Response Guidebook. Although this system is not designed for project management, it does help companies maintain better information and control over hazardous materials used. Companies may avert some contingent costs and less-tangible costs by using software systems such as these, but the costs and benefits cannot be quantified.

User support

The company currently gives users the option to test and use the actual software through a full working program demo for a 30 day trial period. The company offers consulting support, on- and off-site training, toll-free telephone support (9.00 - 6.00 PM Eastern), and annual maintenance programs. The price of the software includes free maintenance and support for one year.

Limitations

The system is limited to hazardous waste management, and may not consider the full range of environmental costs a company may incur. It does not consider social or other external impacts, and no custom reports can be generated.

The current system considers only a limited variety of costs with some options for Customization. Users cannot detail these cost components. The existing cost fields are only for some hidden costs related waste management. Other conventional, contingency, or less tangible costs cannot be estimated or detailed. There is no feature for maintaining historical cost records, as related to quantities and types of materials managed. The system cannot integrate with company accounting systems, resulting in managing these costs separately, rather than integrating them with other company costs. The system is not designed for, and has no modules for, project management. Like most other waste management and tracking tools, the primary objective is to facilitate compliance, not to manage and minimize costs. However, if this area were developed, pollution prevention and other environmental initiatives would be encouraged as a result of better cost information.

Basis for evaluation: Based on telephone communication and information received from Terralink. Additional information received from Rich (1995).

Contact information: Commercially available
Terralink Software Systems Inc. 500 W. Cummings Park, Suite 3150
Woburn, MA 01801
Lisa Naglieri 800-343-5143 617-937-6060 617-937-5577 (FAX)

Docuwaste for Hazardous Waste and other CHEMTOX systems

Prepared by: CHEMTOX-division of Resource Consultants, Inc.

Prepared for: Public purchase

Application: Tracking and reporting of hazardous waste. The system has some predefined fields where waste disposal cost information can be entered. The total costs are broken down to an extent. As yet, the system has no components for project management. There is no feature for maintaining historical cost records.

Elements of control:

Cost control
Resource control
Estimating control
Schedule control
Scope control
Risk control

Development date and updates: CHEMTOX database was developed - 1984 and updated quarterly since. All other modules are updated periodically as well.

Public availability: Available

Purpose and current use: CHEMTOX has installations in over 600 corporations, both in the U.S and abroad. Docuwaste is designed for generators, transporters, and disposal facilities.

Cost information: *Single-user license* CHEMTOX databases $2,000 MSDS Management module: $3,000 Chemical Inventory, Process and SARA 312 Reporting: $6,000 SARA 313 Reporting: $4,100 Docuwaste for Haz. Waste.

The purchase price for multiuser network license is $1,000 per additional user per application. There are discounts offered for multiple unit purchases. These are January 1995 prices, which are subject to change without notice. All prices are for Windows versions.

System requirements: The system can support local PC environments as well as true client/server installations in large corporations. As a stand-alone system, the minimum system requirements for CHEMTOX database are: 25 Me hard disk ,4.0 MB RAM
· CD-ROM drive (optional)

The minimum system requirements for a stand-alone version of MSDS Access are: IBM-compatible PC (386 SX or higher) MS DOS 3.1 5 MB hard disk 2.0 MB RAM.

As a stand-alone system, the minimum system requirements for Docuwaste are: IBM-compatible 286 SX PC MS DOS 3.1 ; 2 MB hard disk ; 640 K RAM.

Software Summary

Size and complexity of projects: Not applicable: the system is not designed for project management. It is meant to manage and control areas of regular company activities. Since the system is modular in format, users can size the system to suit their requirements.

Other compatible systems offered: System is offered in modular format.

Nonquantifiable information: Not applicable; the system is not designed for project management. In the SARA 313 reporting module, electronic notes can be attached to each release result.

The CHEMTOX system is Windows based with a graphical user interface and supports Microsoft's ODBC architecture. The different modules can assist in tracking regulated and hazardous waste, maintaining chemical inventories, reporting compliance issues, and maintaining Material Safety Data Sheets (MSDS) and employee health and safety records.

The CHEMTOX regulated chemicals database is an electronic library containing information on over 10,000 chemicals and hazardous substances. Each substance record has 240 data fields that users can search individually or manipulate for custom reporting. Data include physical properties, reactivity data, exposure limits, transportation requirements, etc. Users may also add their own unique lists of chemicals for custom searches. Chemical names can be searched in a variety of ways such as searching by chemical name, CAS#, DOT ID. RCRA ID. etc. The database can be linked with other modules in the CHEMTOX system, such as SARA reporting, or Docuwaste. The database is also available on CD-ROM or as data elements on electronic tape. The information in the database has a variety of uses, but the system itself does not assist in making decisions for material choices (e.g., choosing between different chemicals to be used in a project). The database does not contain any cost information.

Docuwaste is designed for cradle-to-grave tracking of hazardous waste. Inventories can be tracked by waste description, storage locations, inspections, movement histories, disposal, and costs. Profile information is managed for generators, transporters, and disposal facilities. The detailed waste descriptions include the source of the waste. It has built-in files on treatment codes, hazard classes, container types, etc. Waste profiles can be retrieved from the CHEMTOX database. It facilitates the generation of EPA and state-specific manifests and container labels (EPA, California, or New Jersey). State-specific modules are offered to meet users' specific needs. Year-end totals are provided for completing SARA 313 TRI or Form R reports. It computes year end totals for hazardous waste generator reports. Other CHEMTOX products include MSDS management, SARA reporting, and chemical inventory. All these modules can be bought as an integrated system called MSDS Access. The MSDS management module facilitates collecting, organizing and managing, reporting, and archiving MSDS information. The chemical inventory module calculates threshold amounts, used in compliance reporting, from information retrieved from the MSDS management module. It automates SARA hazard categorization. It allows users to track chemicals on several levels based on their requirements (building, location, process) and also combines like ingredients across a facility. The electronic notes tied to release results in the SARA 313 reporting module can be stored for future reference, allowing users to record information on pollution prevention or waste minimization. The worker right-to-know module can be used in a read- only format (and printing) to access MSDS information from the MSDS management module. Other modules are the CHEMTOX employee module for industrial hygiene, training, and safety; the CHEMTOX DOT system for compliance with DOT'S HM-181; and the CHEMTOX label system.

Life-cycle stages covered

Raw material acquisition
Manufacturing stage
Use/reuse/maintenance
Recycle/waste management

Life-cycle stages covered may vary by module. Docuwaste is a cradle-to-grave waste tracking system, which means it covers the waste management stage of the life cycle. It covers the manufacturing process to the extent that it could track the process responsible for waste generation. It primarily tracks the waste once it has already been generated. The chemical inventory module can also allow chemical tracking in different life-cycle stages, primarily chemical use and reuse in a facility. When the modules are used together, use of particular chemicals could be linked to the generation of different forms of waste. However, Docuwaste only details the chemical composition of the waste and the source. It has no feature for proactively reducing the amount of waste being generated or reducing costs associated with environmental and waste management. These life-cycle stages are not necessarily covered in the case of each individual chemical, material, or product. For example, if a raw material is used up in a manufactured product and does not end up as waste in the facility, the waste management stage will not be covered. Also, the products manufactured in the manufacturing stage are not covered (if at all) in all their life-cycle stages either. The life-cycle stages are not kept distinct within the system.

Type of costs considered

Conventional
Potentially hidden
Contingent
External

Docuwaste considers only hidden costs. It can calculate which waste is produced in the greatest amount with the greatest processing cost. Cost information can be entered for waste descriptions and/or containers in Docuwaste. The fields include disposal cost, freight cost, laboratory analysis, recycling, tax, and other costs. Thus, external and contingent costs cannot be estimated or included. The actual costs (capital, labor and material costs, etc.) associated with environmental management and tracking are not recorded. No additional fields may be added.

Thus, the full costs of using a material or product in a facility, including all these cost categories, cannot be estimated. Also, all conventional and hidden costs are not tracked and entered into the system. The costs entered cannot be collated to form a historical cost library. This could be a useful feature for assisting in cost estimation for projects, because hidden cost estimates cannot be found in most (if any) commercial cost databases.

Method of cost estimation

The system is not designed for cost estimation.

Generation of financial indicators

Net present value (NPV)
Payback period
Internal rate of return (IRR)
Benefits cost ratio
Other

Ability to Include Environmental Costs

User friendliness and flexibility

The system is available in modules that users can add to based on their requirements. It has pull-down menus, pop- up windows, an optional mouse, and help screens. All modules have some features to make the system more user friendly. These include a variety of search options in the CHEMTOX database, the automatic transfer of data from the MSDS management module for SARA reporting, and flags for land ban chemicals in Docuwaste. The chemical tracking module allows tracking on several levels according to user requirements. A relational report writer can be added to the system to produce custom reports.

User support

CHEMTOX also offers a variety of professional services such as chemical inventory survey, MSDS evaluation, data entry, and regular updating of system databases to keep abreast of changing regulations and new scientific knowledge. All product purchases include toll-free support and updates in the first year. After the first year, the maintenance contract can be renewed for a fee. An enhanced customer support service is also available. Training sessions are available both on-site and off-site at Nashville (TN).

Limitations

The screens and formats in the CHEMTOX modules cannot be changed by users. The system considers only a limited variety of costs with no options for additional fields to be added by users. The costs included in the forms are for waste types and containers. Other conventional, contingency, or less-tangible costs cannot be estimated or detailed. The software has no feature for maintaining historical cost records, as related to different activities, quantities, and types of materials managed. The system cannot integrate with company accounting systems. This can result in these costs being managed separately, rather than being integrated with other company costs. The system is not designed for, and has no modules for, project management. Like most other waste management and tracking tools, the primary objective is to facilitate compliance, not to manage and minimize costs. However, if this area were developed, pollution prevention and other environmental initiatives would be encouraged as a result of better cost information.

Basis for evaluation: Based on telephone communication with CHEMTOX representative and information package received.

Contact information: Commercially available
CHEMTOX
7121 Cross Roads Boulevard PO. Box 1848 Brentwood, TN 37024-1848
Jeanie Mackey (Business Development Support Services)
800-338-2815 615-373-5040 615-370-4339 (FAX)

ENFLEX Data

Prepared by: Logical Data Solutions

Prepared for: Public purchase

Application: Database management system for environmental and hazardous material management. The system has some fields where cost information can be entered. Some open fields may be designated as cost fields. There is no feature for maintaining historical cost records. As yet, the system has no components for project management.

Elements of control:

Cost control
Resource control
Estimating control
Schedule control
Scope control
Risk control

Development date and updates: Manufacturing date: 1985. On an average, ENFLEX data is updated once a year. The current release is version 5.01, released in March, 1995. A Windows version of the product is expected in October, 1995. ENFLEX info (manufactured by another company) is updated once every month.

Public availability: Available

Purpose and current use: Not designed for a specific industry. The system is primarily being used in the aerospace, utilities, and discrete manufacturing industries.

Cost information: The average cost of each module in ENFLEX data is $2500. The base system, the facility module, costs $1000.

System requirements: The current version can be operated on a PC or a LAN. Hardware platform/operating systems that ENFLEX is available for include IBM/MVS/TSO, IBM/VM/CMS, IBM/RS6000/AIX, PC/LAN/DOS, DEC/VAX/VMS, and SUN/UNIX. Supported networks are Novell and Windows NT.

Software Summary

Size and complexity of projects: Not applicable; the system is not designed for project management. It is meant to manage and control areas in regular company activities. Since the system is modular in design, users could size it according to their requirements. The vendors say the product is being used in both small and large operations.

Other compatible systems offered: System has 17 modules. ENFLEX Data has a supporting product called ENFLEX info.

There is a limited amount of integration between the two systems.

Nonquantifiable information: Not applicable; the system is not designed for project management. Nearly all the modules of ENFLEX data have a notes section.

ENFLEX Data is a 17-module system for environmental hazardous waste management, which includes chemical inventory and SARA reporting. At the moment, ENFLEX info, is linked with ENFLEX Data in a limited manner. The 17 modules of ENFLEX Data are as follows:

- **Facility Module:** Required module for storing information on processes and locations.

· **SARA Chemical Inventory:** Storage of chemical information and amounts of Extremely Hazardous Substances (EHS) and other chemicals used. Tierl/ll reports

· **313 Form R:** Calculation of 313 chemical releases, reporting (313 Form R).

· **MSDS:** Automated MSDS database. Compliance with HCS, HAZCOM.

· **Chemical Approval:** Formal approval process for chemical use: Requester, Administrator, and Reviewer. Automated questionnaire.

· **Tank Management:** Inventory of all above and below ground tanks, including historical information.

· **PCB Equipment:** Information on all PCB contaminated equipment. Tracking of inspection results, retrofilling process, and preparation of annual TSCA inventories.

· **Container Tracking:** Tracking of all on-site wastes regardless of containers used. Costs can be allocated back to generating location.

· **Manifest:** Multi-part manifests for waste tracking, including transportation and
disposal costs.

· **Air Emissions:** Emissions monitoring for stacks and other sources. Emission history by source or parameter, using built-in graphical or statistical analysis.

· **LDAR:** Tracking of testing and inspection information for equipment that
releases fugitive emissions. SARA 313 TCRR.

· **Water Reporting:** Monitoring water discharge points. Discharge Monitoring Report (DMR).

· **Permit Tracking:** Creating, storing, and monitoring permit life-cycles, for any permit type.

· **Ground Water Monitoring:** Information and tracking water quality in ground water wells.

· **Auditing:** Audit preparation, storage of findings, costs, and follow-up plans.

· **Environmental Events:** Detailed comments for all facility environmental events.

· **Calendar:** Highlights key commitment dates for environmental personnel.

The program has over ISO pre-programmed reports. Report Customization is not an option in the original system, although modification is possible through specific service requests.

ENFLEX Data has a supporting product called ENFLEX info,, which is a database that contains information on hazardous chemicals and regulations. Users can subscribe to ENFLEX info, to receive periodic updates that incorporate new information and regulatory changes.

The system does not currently have a risk assessment component, but the company is in the process of developing a new module for industrial hygiene, which will include risk assessment. Health and safety is not covered in any other module. The system assists users in product selection by considering the potential dangers in selecting one material over another. This application can be used in projects to do qualitative assessments for choosing between different materials based on their environment-related merits and demerits.

Life-cycle stages covered

Raw material acquisition
Manufacturing stage
Use/reuse/maintenance
Recycle/waste management

Life-cycle stage covered may vary by module. For example, the module on container tracking considers information on the waste management stage as the waste material is tracked from the time it is generated or enters a facility to when it is finally disposed of. Use/reuse is covered to the extent that raw material use in a facility is covered. The actual manufacturing process or products of manufacturing are not tracked. However, these life-cycle stages are not necessarily covered in the case of each individual chemical, material, or product. They may not be linked in the sense that the effects of source reduction or process changes on wastes produced and costs cannot be quantitatively analyzed. Also, the products (not wastes) manufactured in the manufacturing stage are not covered (if at all) in all their life-cycle stages either. The life-cycle stages are not kept distinct within the system.

Type of costs considered

Conventional
Potentially hidden
Contingent
External

The software contains cost fields in which users can enter information on various hidden costs. For example, the waste management module has 5 cost fields: taxes, demurrage, transportation, disposal, and other. There are also cost sections for spills and inspection (cleanup costs) and audits (repair costs). Users have a limited amount of flexibility in defining fields and most screens allow users to define 2 to 4 fields (the system as a whole has about 80 open fields). This may allow users to enter some information on additional cost categories. For example, users could enter information on labor or administrative costs involved in performing a certain activity.

The software does not attempt to put a dollar value on contingent or external costs. All other hidden and conventional costs associated with use and reuse of materials, environmental management (e.g., capital, labor), and compliance (e.g., health and safety, administrative costs) are not included in default formats. Thus, all conventional and hidden costs are not tracked and entered into the system. The costs entered cannot be collated to form a historical cost library to be used for cost estimation purposes. This could be a useful feature for assisting in cost estimation for projects, because hidden cost estimates cannot be found in most (if any) commercial cost databases.

Method of cost estimation

The system does not do any type of cost estimation.

Generation of financial indicators

Net present value (NPV)
Payback period
Internal rate of return (IRR)
Benefits cost ratio
Other

Ability to Include Environmental Costs

User friendliness and flexibility

There is a limited amount of flexibility in entering user-defined fields. The modular design of the system allows users to size their systems as related to their operations and requirements. The product selection feature can be useful in project management for qualitative assessments of some material or product choices.

User support

The company offers implementation assistance, assistance with custom reporting, on-site training, and toll-free telephone hotline support. The company also brings out a newsletter and annual ENFLEX Users

Group meetings are organized. No demonstration disks are provided. However, remote demos (enabling software provided) and live on- site demos can be done.

Limitations

The system does not consider all the costs associated with environmental management in a company. The costs entered are hidden costs related to particular activities. Other conventional, contingency, or less-tangible costs cannot be estimated, but can be entered to a limited extent using open fields. There is no feature for maintaining historical cost records, as related to different activities, quantities, and types of materials managed. The system cannot integrate with company accounting systems, resulting in managing these costs separately, rather than integrating them with other company costs. The system is not designed for, and has no modules for, project management. Like most other environmental management tools, the primary objective is to facilitate compliance, not to manage and minimize costs. However, if this area were developed, pollution prevention and other environmental initiatives would be encouraged as a result of better cost information.

Basis for evaluation: Based on information received from Logical Data Solutions.

Contact information: Commercially available
Logical Data Solutions 6825 Capri Place Bethesda, MD 20817
301-365-4336
Bill Hope 419-886-4162 419-886-4172 (FAX)
For ENFLEX info: 610-594-4400

G2 Real Time Expert System, GOA and Neur-on-line

Prepared by: Gensym Corporation

Prepared for: Public purchase

Application: Tools for real-time monitoring and analysis of processes. The greater efficiency afforded as a result of real-time monitoring will result in lower costs, although the systems are not designed to include any cost information. However, cost information could be incorporated into the systems (especially in the off-line applications). The systems have numerous environmental applications, including real-time process and environmental monitoring, identification and diagnosis of problems, and solutions. For example, G2 and GDA have been used for continuous emission monitoring for NOx from utility boilers, real-time monitoring of process and pollution control equipment, tracking waste content and burning, and modeling combustion processes to predict and reduce emission levels and optimize processes. In design applications, the systems may be used to design systems for the lowest life-cycle costs. The Batch Design Kit is an example of such an application.

Elements of control:

Cost control
Resource control
Estimating control
Schedule control
Scope control
Risk control

Development date and updates: Gensym Corp. was founded in 1986. G2 Version 4.0 was released in July, 1995. This is usually updated once a year. GDA version 2.0 was released in February, 1995. Neur-on-line version 1.0 was released in February, 1994. These two products are less frequently updated (in comparison to G2).

Public availability: Available

Purpose and current use: Gensym markets its products to companies with automated processes, government agencies responsible for environmental monitoring and remediation, agencies responsible for traffic control, etc. They have sold their products in over 30 industry sectors in over 30 different countries. The industries include process industries, discrete manufacturing, telecommunications, environmental, power, and aerospace. They have sold over 3,000 product licenses.

Cost information: The range of prices for the products are $6,000 to $37,000. The prices vary with the options chosen and also differ between the on-line and off-line versions.

System requirements: G2 Diagnostic Assistant (GDA): Offered on UNIX workstations from DEC, IBM, HP, and SUN, as well as DEC VAX systems. This requires:

· release 3.0 or higher of the G2 real-time expert system
· 32 MB of memory is recommended for a full GDA development system.

Another system, Neur-On-Line, is additionally offered in Alpha AXP systems. The system requirements are the same as for GDA. 48 MB of memory is required for use with GDA.

Software Summary

Size and complexity of projects: The G2 systems seem more suitable for large, complex operations.

Other compatible systems offered Gensym brings out three different packages called G2 Real-Time Expert System, G2 Diagnostic Assistant, and Neur-On-Line.

Nonquantifiable information: Information on any note-writing capability is not available.

Real-time expert systems are on-line knowledge-based systems that combine analytical process models with conventional process control and heuristics to judge and interpret sensory data, while reasoning about the past, present, and future to assess on-going developments and plan appropriate actions. G2 real-time systems can be applied to address the following environmental areas:

- solid waste, wastewater, pretreatment, drinking water, and emissions management
- TSCA, Clean Air Act, and other State and Federal regulations
- cost management
- waste minimization

Real-time expert systems have been used proactively to minimize leaks, emissions, and waste, as well as to improve product quality, production yield, and energy consumption. They can assist in monitoring for regulatory compliance and identification and diagnosis of unexpected process changes. A real-time system can assist in waste minimization by using ingredient information for a compound to suggest production alternatives and methods to minimize wastes and production costs. It can also do mass-balance calculations.

The G2 Real Time Expert System is used to monitor complex industrial processes and to assist in problem identification. It combines rule and model-based expert system reasoning, real-time execution, user interface graphics, and monitoring system and database interface capabilities.

The G2 diagnostic assistant (GDA) can be used along with G2 and is a graphical, object-oriented decision-making tool. GDA contains predefined graphical blocks that represent information fed into the system through user- or computer-generated input. This information is used to interpret and diagnose real-time data received from controllers and sensors. GDA can be used for problem diagnosis and decision support. Users can also create their own applications within G2. Besides identifying problems after they have occurred, problems can also be predicted using statistical techniques. It can be used for prototyping, modeling, and simulation.

Neur-on-line is designed as a visual, object-oriented tool for building neural networks to monitor complex processes that are difficult to monitor or control with conventional modeling tools or rule-based systems. Neur-on-line has to be integrated with G2 (also GDA), which will perform tasks such as filtering input data and taking actions on the resulting outputs from the neural net.

There are on-line and off-line versions of the systems. A tool kit for modeling applications included in the packages allows users to modify the systems as required.

The Emission Reduction Research Center (ERRC), Massachusetts Institute of Technology (MIT), and Gensym are working on designing a **Batch Design Kit** using the G2 Real Time Expert System. This project was funded by a consortium of nine leading pharmaceutical and chemical companies, along with the US Army, New Jersey Commission on Science Technology, and the USEPA. The goal was to create a software toolkit for use in pharmaceutical and specialty chemical plants to reduce their environmental impacts (enable 'Zero Avoidable Pollution') by optimizing batch process design to minimize pollution. The Kit functions in a Computer Aided Design (CAD) environment for the design of batch processes with ecological constraints. Fed a list of ingredients for a given compound, the program will advise users of different production alternatives and help pinpoint methods that minimize wastes and production costs by selecting the most appropriate reactions: solvents, catalysts, diluents, etc.; reactors; separation and purification processes: and treatment units. General features of the Batch Design Kit include the following:

· Covers the complete life-cycle of the manufacturing process, beginning from the chemical reactions, through pilot-scale tests, to the industrial scale plant, and the plant's operation, revamp, and final shutdown.
· Integrates diverse databases including reactions, materials and their physical properties, equipment, models, regulations, and plant facilities, etc.
· Intuitive in its use, with no programming requirements, intelligent graphic interfaces, and a high level language for human-machine interaction.
· Operate in a distributed client-server architecture to allow access and parallel work by different groups of people (chemists, process developers, process engineers...).
· Utilize existing commercial products, e.g., ASPEN Plus, AUTOCAD.
Idealized designs can be analyzed by enforcing environmental constraints, and more practical designs can be evolved by the guided relaxation of the constraints. Operating and capital costs can be allocated through integration with external databases. It is estimated that the kit would help companies save millions of dollars in non-compliance fines, waste product, and raw materials.

Life-cycle stages covered

Raw material acquisition
Manufacturing stage
Use/reuse/maintenance
Recycle/waste management

Real-time systems can be used to address any part of the life cycle. Thus in real-time applications, it will typically cover stages within one facility that can be monitored. All life-cycle stages will not be addressed in one application. However, in off-line applications, the modeling capability could be used to include information on all life-cycle stages. For example, expert systems could be used in designing applications to help in selecting materials with optimal life cycles, while determining things like reusability of parts, ease of disassembly, best use of materials, etc. However, the system is not preset or designed for this purpose.

Type of costs considered

Conventional
Potentially hidden
Contingent
External

The system is not designed to include cost information, but users could possibly adapt the system to include some variable costs through relationship models. However, users would need to develop estimation methods, such as costs related to material use (and other real-time data),= or external costs related to emissions on their own. The systems can be used to control some cost drivers by maximizing efficiencies. For example, predicting future problems such as leaks or emissions could reduce contingent and external costs (without actually estimating or recording them). Real-time data will allow quicker identification and solution of problems and possibly less damage. In off-line versions, cost information could be incorporated in system models. Since the systems can be modified according to user requirements, any cost information, if available, could be incorporated. However, the system is not designed for estimation or resource control and therefore cannot assist in cost estimation.

Method of cost estimation

The system is not designed for cost estimation.

Generation of financial indicators

Net present value (NPV)
Payback period
Internal rate of return (IRR)
Benefits cost ratio
Other

Ability to Include Environmental Costs

User friendliness and flexibility

Gensym's Telewindows application provides a multiuser client/server environment that allows multiple users to share G2 applications. The bridge products enable plant-wide integration through standard interfaces to popular database and control systems (PLC, DCS). User-defined blocks can be created in GDA. Users can build and maintain applications without needing to do conventional programming. G2 can import bitmapped images and create trend charts for displaying real-time data. G2's internal clock works at subsecond levels for scheduling G2 events and keeping data histories at subsecond time intervals. All systems have graphical user interfaces with the option to use MS Windows or MOTIF style interfaces. The new tool kit for modeling allows users to develop and modify the system application as required.

User support

Gensym provides consulting services to its customers, including project prototyping, application engineering, project management, and customized training. Systems integration services include system architecture and data integration services. No demonstration disks are available. However, remote and on-site demos can be organized.

Limitations

These systems are primarily designed to collect and process real-time data from automatic processes; therefore, they are unsuitable for small, nonautomated operations. The software is not designed as a cost-estimating system or to include cost data. It is meant primarily to be used to handle technical aspects of project planning and monitoring but not other nontechnical (resource control, administration, management) elements.

Basis for evaluation: Based on telephone communication with Gensym representative and information package received.

Contact information: Commercially available
Gensym Corporation 125 Cambridge Park Drive, Cambridge, MA 02140
Joanne Prairie/Betsy Gartner/Steve Fraleigh
617-547-2500 617-547-1962 (FAX)

Gentrax, and other Wixel products Prepared by: Wixel,Inc.

Prepared for: Public purchase

Application: Waste management and tracking. Users can enter different types of cost information (hidden and conventional) into the system, depending on the modules used. It is not clear whether additional user-defined fields may be created. However, historical cost libraries cannot be created for use in project cost estimation. The system has no modules that address project management.

Elements of control:

Cost control
Resource control
Estimating control
Schedule control
Scope control
Risk control

Development date and updates: Introduced - 1989, frequent updates. Version 5.2 of Gentrax was released in July, 1995. A Windows version is expected in 1996.

Public availability: Available

Purpose and current use: Wixel has a client list of close to 500 large and small companies. They have sold their products to the COD, DOE, and EPA. Wixel targets a wide range of users including generators, environmental service companies, and TSDs. They have special capabilities suitable for public utilities and universities. They also target smaller generators and recyclers (dealing with wastes such as composts, plastics).

Cost information: Not available.

System requirements: Designed for IBM-compatible 386 DX PCs.

Capable of operating on mainframes and in the Macintosh environment. Supported networks include Novell, and other PC-based networks.

Software Summary

Size and complexity of projects: Not applicable; system is not designed for project management. The modular design allows users to size the system according to their requirements.

Other compatible systems offered: Wixel offers a variety of packages for users-Gentrax, RecycleTrax, PackTrax, HWTrax, UTrax, TSDTrax, SARATrax, BrokerTrax, and ClassTrax. They are planning to introduce one called CompostTrax. All these packages have been formed using three core programs, which have been coupled with a dozen or so modules.

Nonquantifiable information: Not applicable. Information on any annotation capability is not available.

These packages have been assembled for serving specific user needs. However, users can mix and match modules themselves as well. Areas that these modules cover include the following:

· solid waste, hazardous materials and waste, storm water and runoff, asbestos, and emissions management
· permit tracking, source inventory, MSDS and right to know management, health and safety and training management, risk analysis, property transfers, and audits
· TSCA, Clean Air Act, and other State and Federal regulations · cost management
· recycling, composting
· reporting-including SARA, NPDES, and Form R

· mapping and GIS capability

GenTrax is a waste management system for generating facilities. Gentrax monitors and tracks different parts of users' operations with an option for user-defined parameters. It includes profile and sample tracking, container labeling with optional bar-coding, and warning reports for violations. The billing system allows disposal costs to be charged back to departments responsible for generation. It includes a complete historical database of waste information. Reports include management, shipping, system listings, and cumulative shipping reports (by chemical, waste stream, etc.).

TSDTrax is designed for TSD facilities that conduct various waste processing operations and handling of inbound and outbound wastes. The system has specific portions assigned to management of on-site operations, inbound waste, and outbound wastes. In the on-site operations, material flows are managed through different process operations (including recycling), profiles are managed, manifest information can be electronically transferred, and all site waste is inventoried and monitored. An optional invoicing and accounts receivable module can prepare quotations, invoices, work-orders and include other accounting information.

HWTrax is designed for fixed-State and mobile facilities that deal primarily with household waste. Processes like drop-and-swap, recycling, and hulking for reuse can be tracked. The system includes a chemical database and allows user-defined parameters to add to the areas tracked by the system. Other products include PackTrax, SARATrax, UTrax, UtiliTrax, CompostTrax, and TraxMan.

PackTrax is designed for brokers and environmental service companies dealing with labpacking and fieldwork. **SARATrax** is designed for managing SARA compliance requirements, including inventory, tracking, air emissions, and automatic report generation for Form R, Tier 1, Tier 11, emissions reports, and actual vs. permit limits reports.

Utrax is designed for waste management in universities and research facilities. **UtiliTrax** is made for use in utilities to manage, track, and process RCRA waste and PCB equipment/materials. **CompostTrax** is a management system for private and municipal composters. TraxMan is an information management system for managing paperwork necessary for RCRA compliance. A special PCB module option is available for all standard Wixel programs used by generators.

All systems have features for report generation, container labels, and automatic manifests. The packages include relevant information databases, including information on chemicals, DOT container types, generating process codes, etc. Changing regulations are factored in through frequent enhancements and upgrades.

Life-cycle stages covered

Raw material acquisition
Manufacturing stage
Use/reuse/maintenance
Recycle/waste management

Life-cycle stages covered vary by module. For example, SARATrax tracks individual processes in a production line, calculates mass balances, and tracks emissions. It covers the use and reuse of chemicals. However, the actual products (not by-products and wastes) of a manufacturing process are not tracked; hence this activity is not included as a stage that is covered. GenTrax and TSDTrax will cover recycling/waste management. These different stages cannot necessarily be linked to one another. For instance, users may not be able to link types and quantities of particular chemicals used with a particular quantity of solid waste produced. Gentrax, for instance, does not make this linkage. Thus, users cannot simulate changes in waste quantities and costs as a result of process changes or source reduction. These life-cycle stages are not necessarily covered in the case of each individual chemical, material, or product.

Type of costs considered

Conventional
Potentially hidden
Contingent
External

Where cost information is included, the objective is mostly to facilitate cost tracking or invoicing, rather than cost control and minimization. Gentrax does provide features like disposal cost comparison reports, interdepartmental billing, and others. However, only the costs related to waste disposal are covered. Cost codes in Gentrax include container disposal, consulting charges, miscellaneous charges, profile application, repacking, sample, special handling, and transportation costs. Additional cost codes could possibly be added to cover other cost types. Different types of cost information are covered in different modules. Some, like SARATrax and the PCB module, do not have any cost fields. CompostTrax allows users to track operational costs associated with composting. The optional invoicing module allows users to record costs associated with TSD, transportation, and labor (conventional and hidden costs). Estimates of contingent and external costs could be detailed in this module.

However, the costs of using particular chemicals before they become waste-costs associated with use/reuse (including costs of tracking, monitoring, and other compliance), capital (e.g., pollution prevention equipment) and labor, and other operational costs-are not covered. The costs that are entered cannot be collated to form a historical cost library. Thus, the life-cycle costs associated with using particular chemicals cannot be calculated using the system. If this linkage were made and full costs accounted for, pollution prevention and waste management projects would be encouraged because users would have better information about the full costs of using a particular process, material, or chemical.

Method of cost estimation

The system is not designed for cost estimation.

Generation of financial indicators

Net present value (NPV)
Payback period
Internal rate of return (IRR)
Benefits cost ratio
Other

Information not available. In CompostTrax, various financial reports are available for commercial operations. There is no information on whether any financial indicators are generated.

Ability to Include Environmental Costs

User friendliness and flexibility

Pop-up boxes allow frequently used data to be reused by selecting it from a pop-up display list or table. The quotation/invoicing module can transfer cost information to company accounting systems. Gentrax includes an electronic mail system for communications. The system is menu-driven but has no mouse-support. No custom reports can be generated. Changes in field names (if any) cannot be made directly on screen.

User support

Wixel offers an annual regulatory upgrade and hotline agreement (RUSH) at 20% of the current price of the Wixel software purchased. This upgrade includes implementation support, hotline service, telephone training, and upgrades and documentation. Wixel provides users with slide-show demonstration disks and video tapes to review their products.

Limitations

Besides facilitating compliance, Wixel systems also aim to manage and simplify operations in different areas of waste and environmental management. Various types of cost information can be recorded in different modules. However, cost minimization is not a critical objective. There is no feature for maintaining historical cost records, as related to quantities and types of materials managed. Also, all costs in the four cost categories are not recorded. For example, if a facility generates and treats wastes on site, the system cannot link the life-cycle stages to produce life-cycle costs of using particular materials or processes. The costs of monitoring, permitting, and other compliance are not tracked. Although disposal costs are tracked back to departments of origin, other potentially hidden costs associated with activities such as storage, maintenance, and safety are not recorded. Contingent and less-tangible costs are not estimated or included. The system is not designed for, and has no modules for, project management.

Basis for evaluation: Based on telephone communication with Wixel representative and information package received.

Contact information: Commercially available
Wixel ,7936 E. Arapahoe CT. Suite 3200 Englewood,CO 80112

Ray Richeson (VP-marketing) 303-796-0045 303-796-0043 (FAX)

HAZE

Prepared by: Telstar Associates Inc.

Prepared for: Public purchase

Application: Hazardous material and waste management system. Users can enter environmental cost information into the system. It is not clear what fields exist in the templates and whether user-defined fields may be created. However, historical cost libraries cannot be created for use in project cost estimation. The system has no modules that address project management.

Elements of control:

Cost control
Resource control
Estimating control
Schedule control
Scope control
Risk control

Development date and updates: Introduced - 1992-93. At least one update a year. Telstar has been in this business since 1979. The latest version of HAZE is version 3.05, released on July 14,1995.

Public availability: Available

Purpose and current use: Telstar does not target a particular industry for their product. The company has approximately 42 clients using the product.

Cost information: Each module costs approximately $11,300. Approximate value based on conversation with Telstar.

System requirements: Can be used alone or on a network (LAN/WAN). Supported networks include Novell, Windows NT, and Banyan Vines. The minimum system requirements are:

- IBM-compatible PC (286, or greater)
- MS DOS 3.3 or greater.
- 640K memory
- 40 MB hard disk (recommended for use, program needs 5 MB)
- VGA monitor The system is available on 51/4 inch or 31/2 inch diskettes.

Software Summary

Size and complexity of projects: Not applicable; the system is not designed for project management. It is meant to manage regular company activities. The level of breakdown of the modules (and the price) may not allow the product to be suitable for all sizes and complexities of operations. The system has an unlimited database capability with user-defined table files. The master chemical section has various search options. However, there is no user-defined coding or indexing feature for records. There is a limited amount of Customization (no custom reports, displays, help features).

Other compatible systems offered: The company sells families of products (which can be integrated) for different purposes. These include HazEDisp, HazEStream, and HazETrain.

Nonquantifiable information: Not applicable; the system is not designed for project management. Notes can be entered in the comments area.

The basic HazE system has various companion modules: HazEStream tracks all sampling, HazEDisp tracks all waste, and HazETrain is the training module. All these modules are compatible with each other and are designed to facilitate cradle-to-grave management of hazardous substances through mass balance calculations. The areas covered by all the modules include the following:

- hazardous materials and waste, asbestos, and emissions management
- permit tracking, source inventory, MSDS and right to know management, health and safety management, and training
- TSCA, Clean Air Act, OSHA, and other State and Federal regulations (facilitates compliance with all regulations governing storage, use, and disposal of hazardous substances)
- reporting-including SARA Ill Section 313, and Tier 11 inventory reports

The major components of HazE are the chemical inquiry, master chemical maintenance, reports, table maintenance, utilities, and on-line help. The master chemical database allows users to locate various types of information associated with each substance such as information on components, storage and use, chemical characteristics, and inventory information. The table files allow user-defined table files that can be used to build databases according to users' requirements.

HazEDisp is designed to track and collect disposal and analytical data. It has areas to enter information on waste stream tracking, identification, composition, labeling, and costing. Other functions include waste entity generation tracking, transportation tracking and manifests, and waste disposal and manifest reports. Users have the ability to impose their own thresholds on chemicals.

HazEStream provides a mechanism to track sample collection and analytical data. Using sampling and analytical data, the system does mass balance calculations to calculate the percentage and amount of hazardous constituents discharged into the environment via air, water, soil waste disposal, or product produced. The system allows data to be entered in the form of images.

Life-cycle stages covered

Raw material acquisition
Manufacturing stage
Use/reuse/maintenance
Recycle/waste management

Life-cycle stages covered may vary by module. The product can track hazardous materials from the moment they enter the facility to their final resting place. It can calculate the hazardous constituents discharged into the environment through air, water, soil, or products manufactured. It tracks the amounts of hazardous substances purchased, inventoried, used, and disposed of at various locations. The actual products manufactured are not tracked nor are the manufacturing stages of the life cycle for the various chemicals and materials.

All materials and outputs are not tracked. There is no quantitative linkage between the quantities and types of material used or type of production process and waste generated. Thus, users cannot simulate changes in waste quantities and costs as a result of process changes or source reduction.

Type of costs considered

Conventional
Potentially hidden
Contingent
External

The system aims to track hidden costs associated with hazardous waste management and generates cost summary reports. This feature has been added recently. Costs can be detailed as much as users desire. Thus, other cost categories such as conventional, external, or contingent costs could be added as well. It is not clear whether cost fields can be defined in any area of the program. For instance, if users have a mechanism to quantify costs associated with discharge of hazardous constituents in various media, they can add a cost field to mathematically calculate this based on discharge information.

All conventional and hidden costs are not tracked and entered into the system. External or contingent costs possibly involved in the various life-cycle stages that hazardous materials would go through cannot be estimated using the system. The actual cost involved in the tracking activity is not included. The costs entered in the program cannot be collated to form a historical cost library. This feature could be useful for assisting in cost estimation for projects, because hidden cost estimates cannot be found in most (if any) commercial cost databases.

Method of cost estimation

The system is not designed to assist users in cost estimation.

Generation of financial indicators

Net present value (NPV)
Payback period

Internal rate of return (IRR)
Benefits cost ratio
Other

Ability to Include Environmental Costs

User friendliness and flexibility

The system is PC based and menu driven and has an on-line help feature. It has interfacing capabilities for data exchange with other computer systems (no "seamless" exchange with particular systems). Its import facility allows imports from disks or CD ROM. Custom routines can be created for transferring data from company purchasing, inventory, or bar code system. This feature may allow users to transfer some cost information to company accounting systems, thus facilitating greater integration of environmental and other company costs. Such costs are usually left hidden in company overhead accounts.

HAZE allows unlimited tracking and recording of chemical substances. The system can collect all company notes and data to create a custom disaster plan. The system can compile information previously entered at various sites (up to 999) to generate a centralized corporate report. Quantities entered into the system using various units of measurement are automatically converted into pounds for standard reporting requirements.

User support

A comprehensive user's manual is included in the package, and users can take advantage of toll-free support. Telstar also provides computer consulting and system support. Users can request services for specialized programming and on-site training. Slide-show demo disks and limited-time product trial copies are available.

Limitations

Although modular in design, the system modules are not broken down significantly. The system does not have the popular Windows structure or mouse support. The system comes with a full suite of reports but does not allow Customization of reports. The system is restricted to hazardous substances and waste. It may not cover all the areas of concern related to environmental management (employee health and safety, nonhazardous substances, pollution prevention, etc.) in a company.

Although users can enter cost information in the system, the costing feature is not given as much attention (not mentioned in the information package). The system has existing cost fields for some hidden costs that users can detail. However, it is not designed to take in detailed cost information on all cost categories that could be affected as a result of various activities. Other conventional, contingency, or less-tangible costs cannot be estimated using the system. There is no feature for maintaining historical cost records, as related to various activities and quantities and types of materials managed. The system is not designed for, and has no modules for, project management. Like most other waste management and tracking tools, the primary objective is to facilitate compliance, not to manage and minimize costs. However, if this area were further developed, pollution prevention and other environmental initiatives would be encouraged as a result of better cost information.

Basis for evaluation: Based on telephone communication with Telstar representative and information package received.

Contact information: Commercially available
Telstar Associates Inc. TO. Box 1642 Boise, ID 83701
Jose M. Tellaria, Jr. (President/CEO) Jacques S. Dashnow (Marketing

and Product Support Manager)
208-343-3894 208-385-9168 (FAX)

HAZMIN, and other LOGICAL products

Prepared by: Logical Technology Inc. (LOGICAL)

Prepared for: Public purchase

Application: Hazardous and nonhazardous material and waste management and tracking. The system has some fields where hidden cost information can be entered. No other cost information can be entered (other than in text portions), and the system cannot be modified by users to enter other cost information. The environmental costs entered cannot be detailed. There is no feature for maintaining historical cost records. As yet, the system has no components for project management.

Elements of control:

Cost control
Resource control
Estimating control
Schedule control
Scope control
Risk control

Development date and updates: Introduced - 1985-86. At least two updates a year.

Public availability: Available

Purpose and current use: The software is designed to be suitable for a wide range of companies. It is currently being used in labs, machine shops, utility companies, manufacturing companies, chemical companies, hospitals, and other industries. Currently, there are approximately 500 to 600 sites using the software.

Cost information: Product Single user Site license
HAZMIN $7,800 $22,800 LogiTrac $2,950 $12,950 LogiTrac JR $1,995 $7,900 MSDS SOLUTION $995 $5;500 CCM $695 $4,000 MSDS Image $950 $1,925
These are 1995 prices for single-user systems and site licenses. The company also quotes prices for five users and ten user networks and other options, which are not included here.

System requirements: Can be used alone or on an network. System requirements for a single workstation are:

- DOS 3.1 (or greater)
- 100% IBM-compatible (386 or greater) 2 MB RAM
- one LOO MB hard disk (minimum)
- one floppy drive.

Data storage requirements will differ. Supported networks include different versions of Novell, Banyan Vines, and IBM Token Ring or PC network program.

Software Summary

Size and complexity of projects: Not applicable; the systems are not designed for project management. They are meant to manage and control areas of regular company activities. The systems do have various options for Customization. Users can enter additional information in the Directory or note fields. However, users cannot change the basic format and approach (three documents in HAZMIN) and field names. This may limit users' ability to incorporate the uniqueness or complexities of particular operations.

Other compatible systems offered: Logi Trac, MSDS SOLUTION (TMS), Chemical Compliance Monitor (CCM). All these systems can be integrated (along with HAZMIN) or purchased separately. TMS and LogiTrac can be upgraded to HAZMIN.

Nonquantifiable information: Not applicable; the system is not designed for project management. Notes can be entered in the text portion.

Logical offers four major products: Chemical Compliance Monitor (CCM), TMS (the MSDS SOLUTION), LogiTrac, and HAZMIN. The areas covered by all the products include the following:

· solid waste, hazardous materials and waste, asbestos emissions
· permit tracking, source inventory, MSDS and right to know management, health and safety management, and training
· TSCA, OSHA, and other State and Federal regulations
· reporting-including SARA reports

The Chemical Compliance Monitor (CAM) is a compilation of chemical regulatory and advisory lists and data derived from selected U.S. government, international, State, and other sources. CAM is available as a stand-alone system or as an option with TMS, LogiTrac, or HAZMIN. CCM is available either as a subscription service or as a one-time license. Features of CCM include the following:

· contains over 30,000 chemical names from a variety of sources · determines which Federal regulations apply to chemicals
· finds OSHA and American Congress of Governmental and

Industrial Hygienists (ACGIH) exposure limits, reports quantities, and identifies human and animal carcinogens.
· includes an extensive bibliography for each chemical.

TMS, the MSDS SOLUTION, is a menu-driven package designed for managing Material Safety Data Sheets (MSDSs) and complying with right to know requirements through electronic distribution. TMS is completely upgradable to HAZMIN. Features of TMS include the following:

· standardized format that includes fields for all common MSDS physical and chemical data and components. This format assists in OSHA training by minimizing variance in MSDS formats.
· custom report-writing feature and any data fields, components, and text can be used to create a custom report.
· unlimited number of locations can be identified for each MSDS showing use, transfer, etc.
· status system can be used to differentiate MSDSs for current/approved materials from others in the system.
· password security system that can limit access in various ways

LogiTrac is a material tracking system, which can be used along with TMS and CCM to track all hazardous materials and waste from purchase to disposal. When LogiTrac is linked with CCM and TMS, MSDSs can be cross-referenced with regulatory and advisory lists. All MSDS fields and text can be queried. Features of LogiTrac include the following:

· records necessary for regulatory compliance and reporting
· completely upgradable to HAZMIN and compatible with all other Logical products
· tracking can be customized according to user requirements
· allows users to establish names of storage areas, work areas, disposal streams, and wastes

· custom reporting feature allows queries of all movements and an unlimited number of subqueries
· can be integrated with existing purchasing and distribution reporting systems

HAZMIN, the Hazardous Materials Information Network, combines all the features of LogiTrac with a product review for materials management and compliance with regulations such as SARA reporting, right to know training, and OSHA's 30-year rule for record retention. HAZMIN uses three types of documents: the MSDS, Request, and Notice. MSDSs are entered into HAZMIN using a combination of predetermined fields and user-named free-form text areas. After an MSDS is loaded onto HAZMIN, a request can be made when use of a product is proposed or changed. The request is reviewed (including looking for possible substitutes-nonhazardous or less expensive), and the request is either granted or denied. A notice is written by the committee for each approved hazardous material, which is used in safety and other procedural training. The notice is the principle training document for a specific product. A status system allows currently approved materials to be distinguished from materials under review or other old information. The system can be used to generate custom reports. LOGICAL also offers modeling software (included in HAZMIN and LogiTrac) for Form R threshold and release calculations, disposal stream monitoring, flexible emission standards, and other applications. Models can be based on material/waste as a whole or on a specific component.

LogiTrac, LogiTrac JR, and the MSDS SOLUTION do not contain CCM or MSDS text. The imaging software integrates with existing Logical software and offers additional functions for using MSDS images.

None of the systems described above have any functions for project management. However, users can use them for information to analyze some material options in a qualitative manner, such as choices between different chemicals to be used. The MSDSs or the Chemical Compliance Monitor can provide information on this.

Life-cycle stages covered

Raw material acquisition
Manufacturing stage
Use/reuse/maintenance
Recycle/waste management

Life-cycle stages covered may vary by module. The use of materials can be tracked from initial use to final disposal. These stages are not differentiated in the system. Wastestreams can be tracked back to the process or department that was responsible for generating them. Ideally, requests are reviewed prior to material acquisition and use in the manufacturing process. The actual products of any manufacturing activity cannot be tracked using the system.

All materials and outputs are not tracked. There is no quantitative linkage between the quantities and types of material used or type of production process and waste generated. Thus, users cannot simulate changes in waste quantities and costs as a result of process changes or source reduction.

Type of costs considered

Conventional
Potentially hidden
Contingent
External

Cost information can be entered in the text portion or incorporated in some existing cost fields. Thus, if users have information on some other cost categories, they can enter that in the text portion, but this approach does not facilitate cost control. The only types of cost fields that exist are for entering costs for manifests. Thus all hidden costs associated with compliance are not included. The Requests are designed to additionally pick up external costs associated with company image and relationship costs.

The full costs of using a material or product in a facility, including all these cost categories, cannot be estimated using the system. The costs entered cannot be collated to form a historical cost library. This feature could be useful for assisting in cost estimation for projects, because hidden cost estimates cannot be found in most (if any) commercial cost databases. However, if users do enter cost information (e.g., costs associated with tracking in LogiTrac) they may be able to export the cost information to another database or spreadsheet. This could enable them to separately maintain cost records, which can further be used in project cost estimation.

Method of cost estimation

The system is not designed for cost estimation.

Generation of financial indicators

Net present value (NPV)
Payback period
Internal rate of return (IRR)
Benefits cost ratio
Other

Ability to Include Environmental Costs

User friendliness and flexibility

All the software systems are menu-driven. CCM allows users to search for chemicals using full or partial names, synonyms, or GAS numbers. TMS, LogiTrac, and HAZMIN have custom report-writing features. Some standard reports are already included in the system. LogiTrac allows customized tracking in terms of how closely an item will be tracked and where it is tracked to and from. Since the systems are written around a relational database (Paradox) they have the capability to export and import data from a variety of software systems.

User support

The company provides licensed users with free maintenance and support for a period of 12 months, after which users have the option to purchase the service. The fees for most products include different numbers of days of training with additional training at a fixed price per day. Each product has a different maintenance and support fee associated with it. The support includes toll-free phone support and semi-annual software updates. CCM customers receive four updates per year. Services include a quarterly newsletter, semi-annual user-group meetings, and 24-hour emergency support. LOGICAL also undertakes requests for custom programming, MSDS loading and database maintenance, and consulting for hazardous materials management and for system implementation. The full price of any LOGICAL system can be applied to the purchase of an upgrade.

Limitations

LOGICAL recognizes that material management saves a company money, but they do not give cost management as much importance.

Although LOGICAL customizes systems, they do this only through special requests to LOGICAL. If users make changes in the system themselves, they can no longer receive LOGICAL's support.
The system currently has fields for only limited types of costs. Life-cycle information entered will vary according to the type of material and compliance requirements. For most cases, dynamics outside the company such as social costs will not be estimated or considered. There is no feature for maintaining historical cost records, as related to different activities and quantities and types of materials managed. The system cannot integrate with company accounting systems, which can result in managing these costs separately, rather than integrating them with other company costs. The system is not designed for, and has no modules for, project management. Like most other waste management and tracking tools, the primary objective is to facilitate compliance, not to include cost elements and detailing. However, if this area were developed, pollution prevention and other environmental initiatives would be encouraged as a result of better cost information.

Basis for evaluation: Based on telephone communication with logical representative and information package.

Contact information: Commercially available
Logical Technology 5113 N. Executive Drive Peoria, IL 61614-4868
Connie Cummings (Sales Associate)
800-COMPLY 1 309-689-2900 309-689-2911 (FAX)

Material Inventory Report System

Prepared by: AV Systems

Prepared for: Public purchase

Application: Environmental, materials and waste management. The primary function is compliance with different environmental regulations. The system has some fields where hidden cost information can be entered. No other cost information can be entered (other than in text portions), and users cannot modify the system to enable this. The environmental costs entered cannot be detailed. There is no feature for maintaining historical cost records. As yet, the system has no components for project management.

Elements of control:

Cost control
Resource control
Estimating control
Schedule control
Scope control
Risk control

Development date and updates: The system was developed in 1988. Updates are provided twice a year. The current version is, version 3.31, released in April, 1995.

Public availability: Available

Purpose and current use: The system is intended to have cross-industry applications. It has most commonly been used by manufacturing companies (although also used in other industries).

Cost information: The approximate price of each module (based on conversation with AV Systems representative in April 1995) is $1,500. The purchase price will vary with options chosen.

System requirements: Designed for IBM PCs and PC networks (supports various networks, including Novell). Concurrent user options are provided.

Software Summary

Size and complexity of projects: Not applicable; the system is not designed for project management. It is meant to manage and control regular company activities in different areas. The modular format allows companies to size the system as required.

Other compatible systems offered: Package is modular in design with options for EPA-approved Form R, SARA Title Ill, Hazcom, MSDS, permit, air, water, waste, inventory, and State and local reporting. Modules can be used alone or in integrated forms.

Nonquantifiable information: Not applicable; the system is not designed for project management. No annotation capability. The company brings out packages that address specific areas of environmental management such as MSDS, NPDES, Permit, etc. The Material Inventory Report System (MIRS) contains all the features of the various modules offered.

· solid waste, hazardous materials and waste, storm water and runoff, wastewater, groundwater, pretreatment, drinking water, asbestos, and emissions management, permit tracking, source inventory, MSDS and right to know management, health and safety management, and training
· TSCA, Clean Air Act, and other State and Federal regulations
· cost tracking
· reporting-including SARA, NPDES, and Form R
· pollution prevention

The MIRS module tracks hazardous and toxic chemical information for compliance with EPA SARA Title Ill and includes a regulated chemical database. The MSDS modules are designed for OSHA HAZCOM compliance and allow users to enter, track, update, and archive MSDS information. The Netview module facilitates company-wide access to MSDS information. The Form R and Easy-R modules cover compliance with SARA Title Ill Section 313 toxic release reporting. Easy-R contains multiyear comparison reporting for pollution prevention. The VOC module tracks emissions data and air permits using several methods of VOC computation. The WASTE container and manifest module handles cradle-to-grave waste tracking. RCRA biennial and other reports can be generated using this module. Other modules include INVENTORY for chemical inventory transactions recording, PERMIT for organizing a facility's environmental permits, NPDES for Clean Water Act (CWA) compliance; and LOCAL to cover reporting variations from the EPA.

Life-cycle stages covered

Raw material acquisition
Manufacturing stage
Use/reuse/maintenance
Recycle/waste management

Life-cycle stages covered may vary by module. The waste tracking module tracks only waste materials from the point of generation to final disposal. The INVENTORY and MSDS account for chemical and hazardous material use in a facility. The raw material acquisition stage and the manufacturing of these materials are not covered. Thus, although using different chemicals is covered in a different module, the actual link between quantity of chemical usage or production and waste generation cannot be made. Users cannot simulate changes in waste quantities and costs as a result of process changes or source reduction. However, the WASTE module can link wastes to the process, chemical, or department responsible for creation. The air emissions module does function to link type and quantities of particular chemicals to VOC emissions.

Type of costs considered

Conventional
Potentially hidden
Contingent
External

Only the WASTE module has actual cost fields. It can include hidden costs such as manifest costs, transportation, and disposal costs. These costs can be linked back to the process or department responsible for their creation. There is a field called "other costs" where other costs can be specified. There is no function in the program to maintain historical hidden cost records as related to types and quantities of chemicals used. However, cost data could be exported to Excel or other spreadsheets and database packages, where users could develop cost trends that could be used for estimating purposes.

Also, all conventional, hidden costs, and contingent costs are not tracked and entered into the system. Capital, labor, material, and other costs associated with environmental management in a facility cannot be detailed. Thus, the full costs of using a material or product in a facility, including all these cost categories, cannot be estimated using the system.

Method of cost estimation

The system is not designed for cost estimation.

Generation of financial indicators

Net present value (NPV)

Payback period
Internal rate of return (IRR)
Benefits cost ratio
Other

Ability to Include Environmental Costs

User friendliness and flexibility

The program is menu driven with an optional Windows format. Field and context-specific help is available. Since the program is built in a dBase compatible format, information in the modules can be exported to other software packages such as Excel.

User support

AV Systems offers consulting, training, and maintenance support. On-site training is also available. The maintenance support includes unlimited telephone support and program updates. Slide-show demonstration disks are provided.

Limitations

Life-cycle information entered will vary according to the type of material and compliance requirements. The system currently has fields for only limited types of costs. Dynamics outside the company such as social costs cannot be estimated or considered. There is no feature for maintaining historical cost records, as related to different activities and quantities and types of materials managed. The system cannot integrate with company accounting systems, which can result in managing these costs separately, rather than integrating them other company costs. The system is not designed for, and has no modules for, project management. Like most other waste management and tracking tools, the primary objective is to facilitate compliance, not to manage and minimize costs. However, if this area were developed, pollution prevention and other environmental initiatives would be encouraged as a result of better cost information.

Basis for evaluation: Based on telephone communication and information received from AV Systems in March 1995.

Contact information: Commercially available
AV Systems 4657 Platt Road Ann Arbor, MI 48108-9795
Karen Wilson 313-973-3000 313-677-4480 (FAX)

PlantWare

Prepared by: EnviroMetrics Software Inc.

Prepared for: Public purchase

Application: Multimedia emissions tracking. The systern has the potential to be developed for environmental cost management. It includes existing fields where costs related to treatment, storage, and disposal can be entered and tracked back to departments responsible for waste generation. Information on other costs can be detailed as required by users by adding additional cost fields. Costs due to fugitive emissions can be estimated. As yet, the system has no components for project management. It is possible to record, track, and report historical cost data for environmental activities, including capital, acquisition, waste disposal, etc.

Elements of control:

Cost control
Resource control
Estimating control

Schedule control
Scope control
Risk control

Development date and updates: Envirometrics has been in business since 1991. PlantWare was released in early 1994. Version 1.4 was released in July, 1995.

Public availability: Available

Purpose and current use: EnviroMetrics has a client list of close to 100 customers, primarily in the Chemical, Utility, Pulp and Paper, and Pharmaceutical industries. Their clientele includes names like Amoco, Mobil, Georgia-Pacific, DuPont, and Weyerhauser.

Cost information: The average sale price for the three modules ranges from $12,000 to $15,000, depending on volume purchased. Prices approximate values, based on conversation with Envirometrics representative in April 1995.

System requirements: Companies can choose from a variety of hardware and database engine options, including SQL Server, ORACLE, and SYBASE. The database can be supported by a variety of platforms and operating systems. PlantWare is a true client-server, Windows- based product that can be used on a local area or wide area network. It supports many popular networks such as IPX, NT, Novell, and NETBIOS.

Software Summary

Size and complexity of projects: Not applicable; the system is not designed for project management. It is meant to manage and control areas of regular company activities. The ability to customize the system allows users to modify the system to fit their requirements. However, the system may not handle the full range of environmental issues a company may have to address.

Other compatible systems offered

The system is available in modular or integrated form. Thus, PlantWare can be
integrated with other available modules.

Nonquantifiable information: Not applicable; the system is not designed for project management. Complete annotation is available via user-defined fields.

PlantWare is a Windows-based, client-server software tool for multimedia emissions management. It is designed for managing air, water, and solid waste emissions, using one central relational database. At the heart of the system are tools for mathematically representing facility emissions, using simple or complex expressions. For example, emissions can be modeled as a function of time-dependent variables such as production rates. What-if scenarios can be analyzed by altering values of input variables. Future emissions can also be projected. Features of the system include the following:

· solid waste, hazardous materials and waste, storm water and runoff, wastewater, and emissions management
· permit tracking, source inventory, and training
· Clean Air Act management
· flow editor for graphically linking diagrams sources, control devices, and emission points
· engineering tools for calculating emissions with a library of standard emission equations and tools for custom equations
· standard batch emission equations for tracking emissions from batch chemical facilities

· tools for tracking hazardous and nonhazardous solid waste
· reminders and notes for important dates and activities
· interface with Oracle, Sybase, SQLServer, Ingress, RdB, DB2, and several other SQL database backends
· standard report library (SARA Form R, NPDES permits. Title V permitting, and others) and custom report generator

Envirometrics offers other packages or modules that address MSDS and right to know management, waste minimization, groundwater, drinking water, and those that address State and Federal regulations. They have another product called the Fugitive Emissions Management System (FEMS).

Project managers can use the emissions modeling function and the "what if" capability to predict emissions and related costs as a result of changes in production, chemicals used, etc. PlantWare has the ability to capture, track, and report costs associated with specific chemicals, processes, equipment, etc. Thus, to an extent, it facilitates the comparison of different materials or chemicals based on quantitative criteria. It may be possible to extend this feature to predict costs associated with the use of a process/product for the purpose of project planning.

Life-cycle stages

Raw material acquisition
Manufacturing stage
Use/reuse/maintenance
Recycle/waste management

PtantWare considers all waste management operations and emissions in a facility. It considers waste management up to final disposal for some components such as hazardous materials. All materials and outputs are not tracked. A quantitative linkage can be developed between the quantities and types of material used or type of production process and waste generated. Thus, users can simulate changes in waste quantities and costs as a result of process changes or source reduction. However, this can be done only for emissions management (not for all types of waste).

Type of costs considered

Potentially hidden
Conventional
Contingent
External

PlantWare comes with fields for recording some conventional and hidden costs associated with environmental management within a firm. The waste-tracking module facilitates the allocation of treatment, storage, and disposal costs to business units within a firm. All conventional costs, such as labor, capital, and material costs that companies would need to incur during the course of environmental management activities are not included. However, users have the ability to add custom fields that will enable them to detail costs as required. Proper cost detailing can help facilitate recording information on all cost categories. However, users would need to estimate external and contingent costs on their own. The system will not assist in cost estimation per se, but users can predict costs due to current and projected emissions of particular chemicals. If users can estimate a external (social) cost per unit of emission, this cost category may be detailed in a similar manner. The costs entered cannot be collated to form a historical cost library. However, they can be archived along with other information. Another option can be to export cost details to a separate spreadsheet or to other packages. This feature could be useful for assisting in cost estimation for projects, because hidden cost estimates cannot be found in most (if any) commercial cost databases.

Method of cost estimation

The system is not designed for cost estimation. However, users can estimate costs due to losses from projected emissions by assigning a cost per pound for a chemical. They can use this to help estimate true costs of chemical use as part of projects. For example, if users are considering using one solvent versus another, costs due to fugitive emissions can be factored into the analysis.

Generation of financial indicators

Net present value (NPV)
Payback period
Internal rate of return (IRR)
Benefits cost ratio
Other

No financial indicators are generated, but it may be possible to export cost data to spreadsheets and other packages that do have financial functions.

Ability to Include Environmental Costs

User friendliness and flexibility

The system has options and features that give users the flexibility to modify the system according to their requirements. For example, the Run-Time Data Dictionary can be used to configure and modify any window in the system. Although the primary objective of the software is to facilitate compliance, the additional features such as those for cost management indicate a move toward becoming a corporate environmental management system.

User support

PlantWare provides a demonstration disk of their software upon request. Complete user-support is offered, including unlimited phone support and optional on-site services. Slide-show demonstration disks are available for PlantWare. On-site demos are provided. FEMS has a working demo disk.

Limitations

The main focus of PlantWare is emissions modeling. The existing cost fields can be further detailed. However, contingent and less-tangible costs associated with emissions, disposal mechanisms, etc., are not estimated. This area, if developed, can be a useful feature in quantifying all costs related to using particular chemicals. For example, an equation could be developed that estimates possible future liability costs as a function of fugitive emissions from using a particular chemical. This can help project managers make more informed choices between different materials to use (rather than one-time conventional costs alone). The costs entered into the system cannot be exported to company accounting systems, which can result in managing these environmental costs separately, rather than integrating them with other company costs. The system is not designed for, and has no modules for, project management.

Basis for evaluation: Based on telephone communication with EnviroMetrics representative and information package.

Contact information: Commercially available
EnviroMetrics Software Inc. 92 Read's Way, Suite 103 New Castle, DE 19720
Jim Braselman 302-324-9136 302-324-9138

WASTE DB TM, and Other Envirogenics Products

Prepared by: Envirogenics

Prepared for: Public purchase

Application: Marges hazardous waste from initial disposal to final disposition. Also prints manifests, reports, and container labels. The system has one cost field where manifesting costs can be entered. No other costs are detailed, and custom fields cannot be created. There is no feature for maintaining historical cost records that can be used for estimation purposes. As yet, the system has no components for project management.

Elements of control:

Cost control
Resource control
Estimating control
Schedule control
Scope control
Risk control

Development date and updates: CHEM Master® line of products, including Waste DB™ were introduced in 1985. The current version of CHEMMaster® is version 3.3, released in January, 1995. The most current version of Waste DB TM was updated in the middle of 1994.

Public availability: Available

Purpose and current use: Envirogenics has sold their CHEM Master® systems to almost 1000 clients with names including

DuPont, General Electric, and AT&T.

Cost information: The cost of the system is $950. It is licensed by SOFTTouch Systems Inc.

System requirements: The system requirements are:

· IBM PC/PS2 with a minimum CPU recommendation of 80286 (base)
· PC DOS version 3.3 * 520 KB RAM'
· 0.3 MB initial hard disk space
· CGA/EGA/VGA monitors
· printer that can print 12-pitch.

The system can be upgraded to work on any network that can support DOS.

Software Summary

Size and complexity of projects: Not applicable; the system is not designed for project management. It is meant to manage and control areas of regular company activities, including waste manifesting and tracking. Not meant for very large/complex requirements.

Other compatible systems offered: CHEMMaster®, MSDS Generator™.

Nonquantifiable information: Not applicable; the system is not designed for project management. No notes can be entered.

Waste DB™ is a system in which users can record and track information related to waste and waste management. Users input information on work area, waste material, processing, waste handlers, waste containers, and waste manifests. Waste DB performs functions that include the following:

· tracking waste disposal by manifest number and waste containers from point of origin to destination
· tracking cost of waste by facility, process, or department
· printing manifests, hazard and shipping labels, and generates waste summary reports organized by process, TSDF, and transporter
· codes and DOT hazard descriptions
· user-developed library of information on transporters and TSDFs and other historical information
· generates EPA Biennial Generator report

The system assists users in complying with State and Federal reporting requirements. Waste DB™ has generic Federal templates, and user-designed templates can be used for creating State-specific templates. It prints Uniform Hazardous Waste Manifests in the generic Federal format, or customized to individual state formats. Waste DB™ can also generate cost reports. Data is organized with the objective of minimizing entry of information, and for easy retrieval or long-term recordkeeping.

CHEM Master®, a separate product containing a regulatory database, is not included as part of Waste DB™. It contains hazard information on all regulated chemicals that can be used for waste analysis and hazard evaluation.

Life-cycle stages

Raw material acquisition
Manufacturing stage
Use/reuse/maintenance
Recycle/waste management

The system only considers the life-cycle stages in which waste is created and finally disposed of. It covers only the waste management stage of the life-cycle and does not consider recycling waste. Chemical and material use in a facility is not inventoried or tracked. The system has the ability to track costs of waste by business group or department. Users can assign a process number or work area to each waste container generated, which can provide useful information for pollution prevention initiatives. However, the program itself does not assist in evaluating options for waste minimization or for recording them.

There is no quantitative linkage between the quantities and types of material used or type of production process and waste generated. Thus, users cannot simulate changes in waste quantities and costs as a result of process changes or source reduction.

Type of costs considered

Conventional
Potentially hidden
Contingent
External

Cost information can be recorded by manifest and/or waste container. This is a hidden cost associated with regulatory compliance. This information can be entered as a single figure and cannot be broken down to show cost details. Costs can be tracked and reported by manifest line items, work area codes, range of dates, manifest document numbers, TSDF, and transporter.

All conventional and hidden costs associated with waste management are not tracked and entered into the system. Users cannot estimate or record any other costs, including usual operating costs, contingent, or external costs. Thus, the full costs of using a material or product in a facility, including all these cost categories, cannot be estimated using this system. The costs that are entered cannot be collated to form a historical cost library.

Method of cost estimation

The system does not do any cost estimation.

Generation of financial indicators

Net present value (NPV)
Payback period
Internal rate of return (IRR)
Benefits cost ratio
Other

Ability to Include Environmental Costs

User friendliness and flexibility

Waste DB™ only allows a limited amount of Customization in printing reports, etc. It does not have many of the advanced functions of other, more integrated software systems, but it may be less expensive than them as well. Fields cannot be added or deleted. However, the company does add customized functions (for custom screens and specific functions) upon specific request from clients. It has pop-up windows and an online help system that provides information on every menu and input field, but it does not have a graphical user interface.

User support

The company offers services in the areas of environmental and industrial hygiene consulting, implementation, programming, data entry, data analysis, and training. The system's price includes one year of toll-free user support. However, the program and documentation are provided only with a warranty of 90 days and the licensee assumes all further costs related to any required servicing, repair, or correction. Demonstration disks can be purchased at a cost of $24.95. The accompanying manual provides a detailed description of the software system and its capabilities. Envirogenics also offers services to industry in the areas of hazardous waste management, indoor air quality, environmental consulting, regulatory compliance, and health and safety training.

Limitations

Waste DB™ is not available in modular form and serves very specific functions. This can be a limitation to those who wish to expand the system or integrate it with other environmental management activities (e.g., health and safety) in the firm. The program cannot be modified without written consent from the licenser. Users can do only a limited amount of Customization. There is no mouse support and users will have to familiarize themselves with various key commands to use the system.

The system considers only manifest and waste container costs, and users have no options for adding additional fields. They cannot be detail cost components. Thus, other conventional, contingent, or less-tangible costs cannot be estimated or detailed. It has no feature for separately maintaining historical cost records, as related to different activities, quantities, and types of materials managed, for the specific purpose of use in future project cost estimation (although the Waste DB™ data records can be maintained). The system cannot integrate with company accounting systems, which can result in managing these costs separately, rather than integrating them with other company costs. The system is not designed for, and has no modules for, project management. Like most other waste management and tracking tools, the primary objective is to facilitate compliance, not to manage and minimize costs. However, if this area were developed, pollution prevention and other environmental initiatives would be encouraged as a result of better cost information.

Basis for evaluation: Based on telephone communication with Envirogenics representative and information package including a demonstration disk, received in February 1995. Additional information obtained from Rich (1995).

Contact information: Commercially available
Envirogenics
3812 B Quakerbridge Road, Suite 208 Mercerville, NJ 08619
800-437-7990 609-586-4426 (FAX)

Waste Trail and other Corbus Systems

Prepared by: Corbus (A division of Compliance Management Systems Inc.)
Prepared for: Public purchase

Application: Environmental and safety management. The system has some fields where cost information can be entered. A limited number of environmental costs can be entered, though not detailed. There is no feature for maintaining historical cost records. As yet, the system has no components for project management.

Elements of control:

Cost control
Resource control
Estimating control
Schedule control

Scope control
Risk control

Development date and updates: Introduced in 1994

Public availability: Available

Purpose and current use: TERMS is targeted to the manufacturing industry. Prolab (and Plus) is designed mainly for laboratories. Corbus has clients in many different companies including Proctor and Gamble, Westinghouse, and GTE.

Cost information: The prices (effective September 1994) are as follows:
Software PC LAN TERMS $3,500 $8,000 PROLAB PLUS $5.000 $8,000 MSDS Standard $1,195 $2,000 WASTE TRAIL $995 $1,900

The multiuser price for TERMS is $5,900. These prices are subject to change.

System requirements: System requirements for Prolab are:
· IBM-compatible PC (min. XT or AT preferred)
· 20 MB hard disk space (varies)
· MS-DOS 3.3 (or greater), and
· 640 K RAM.

System requirements for TERMS are:

· MS-DOS 3.3 (or greater),
· 640 K RAM, and
· 8 MB of hard disk space The Windows version of TERMS is to be released in the fall of 1995.

Software Summary

Size and complexity of projects: Not applicable; the system is not designed for project management. It is meant to manage and control areas of regular company activities. Corbus sells families of products; these have a moderate amount of breakdown. Users can integrate most of them. The products differ in the amount of Customization possible, information sharing, and comprehensiveness; thus, they cannot handle all operation uniqueness and complexities.

Other compatible systems offered: Corbus offers software systems called Prolab Plus and TERMS that can be integrated with Waste Trail. They also bring out a program called MSDS Standard.

Nonquantifiable information: Notapplicable; the system is not designed for project management. No note writing capability.

The other software systems-PROLAB Plus, TERMS, MSDS Standard, and Waste Trail-cover the following areas:

- solid waste, hazardous materials and waste, and emissions management
- permit tracking, source inventory, MSDS and right to know management, health and safety
- cost management
- Clean Air Act, OSHA, CERCLA, and other State and Federal regulations
- reporting-including SARA reports and Form R, CERCLA Spill Report

The Environmental Regulation Management System (TERMS) assists plants with MSDS management, including indexing MSDS by name, part number, synonyms, department, location, hazardous ingredients, and suppliers. It automates SARA Title Ill reporting and Clean Air Requirements. It includes HAZCOM for MSDS management and archiving, worker training, and hazardous materials checklist; a complete reference database (Federal and State regulations) and chemical information; and inventory management.

MSDS Standard deals only with storing MSDS records. Its features include ANSI Z400.1 format for all MSDSs, selector index to retrieve specific sections of MSDS, an emergency hot key, MSDS archiving that complies with the OSHA 30-year rule, and a written hazard communication plan. It has a multiple data entry option and options include direct input and electronic data transfer. MSDSs can be organized into specific categories according to user requirements. It also has multiple security levels.

Prolab Plus (and Prolab) has been designed for chemical inventory management in a laboratory setting. Prolab only has inventory, container labeling, and MSDS management modules. Prolab Plus additionally has features that include bar coding, a regulatory database (with pre-programmed chemical lists with pertinent information), requisition and purchasing, incident reporting, and a custom report writer. It also includes a free-form database manager.

Waste Trail is designed to track hazardous waste and includes cradle-to-grave container tracking, generation of waste manifests and profiles, and various pre-programmed waste tracking reports. Waste definitions can be added or edited including CAS#, EPA hazard #, and Hazard class. Cost breakdowns for waste removal can be tracked by department.

Life-cycle stages covered

Raw material acquisition
Manufacturing stage

Use/reuse/maintenance
Recycle/waste management

Life-cycle stages covered may vary by module. Waste Trail allows cradle-to-grave container tracking (from moment of generation in facility to final disposal). Prolab Plus (and Prolab) has been designed for chemical inventory management in a laboratory setting, which covers the use/reuse stage of the life cycle. All materials and outputs are not tracked. There is no quantitative linkage between the quantities and types of material used or type of production process and waste generated. Thus, users cannot simulate changes in waste quantities and costs as a result of process changes or source reduction.

Type of costs considered

Conventional
Potentially hidden
Contingent
External

Waste Trail allows users to track departmental cost breakdowns for waste removal. It is not clear how much the costs can be broken down. Information on contingent and external costs is not estimated or tracked. The costs entered cannot be collated to form a historical cost library. This feature could be useful for assisting in cost estimation for projects, because hidden cost estimates cannot be found in most (if any) commercial cost databases.

All conventional and hidden costs are not tracked and entered into the system. Conventional costs, such as labor, capital, and material costs, that companies would need to incur during the course of environmental management activities cannot be detailed. Thus, the full costs of using a material or product in a facility, including all these cost categories, cannot be estimated.

Method of cost estimation

The system is not designed for cost estimation.

Generation of financial indicators

Net present value (NPV)
Payback period
Internal rate of return (IRR)
Benefits cost ratio
Other

Ability to Include Environmental Costs

User friendliness and flexibility

Waste Trail is menu driven and has many look-up tables. Corbus provides HAZREG, which is a chemical and regulatory database. Data export and import capabilities vary. TERMS can integrate data from other inventory control and purchasing software systems. It can also provide modem interface.

ProlabPlus has features for handling custom reporting (from inventory records, purchasing records and employee incident data) and unique regulatory requirements. Prolab has a menu-driven query system for referencing. It is compatible with TERMS for SARA reporting and with MICAS for MSDS management.

User support

All software comes with documentation, toll-free technical support, updates (free for the first year), and on-site training ($750 per day). The price for maintenance options for TERMS and Prolab Plus are $800/year (PC) and $1 .SOD/year (LAN), with $1,600 for multiuser support TERMS. Corbus also provides MSDS data entry services. Corbus provides training programs for groups (30 employees average) regarding the basic requirements on right-to- know ($1,000). The training is customized to correlate with hazards specific to the company in question.

Corbus also has a team of experts who can provide environmental audit (for environmental and safety issues) services to companies. The price of the comprehensive site audit (one day) is $1,250. Corbus offers users the service to scan their MSDSs, with the option to store them on CDs. Scanned MSDSs are compatible with Corbus TERMS™ and Prolab™ Plus. Working demonstration disks are available.

Limitations

The amount of Customization depends on the product. WasteTrail, for instance, does not have very many features for Customization. TERMS does not have features for custom report writing. Displays, help screens, and work areas cannot be customized. The system is limited to hazardous materials and waste management and does not address the whole gamut of environmental issues that may face a company (multimedia emissions management, nonhazardous waste management, waste minimization, pollution control, etc.). All the modules cannot share information in a comprehensive manner. Cost information and cost management is not covered in all the products. All cost information related to environmental management (costs of permits, compliance, operating costs, etc.) in a company cannot be viewed.

The system considers only a limited variety of costs with some options for users to add additional fields. All conventional, contingent, or less-tangible costs cannot be estimated or detailed. There is no feature for maintaining historical cost records, as related to different activities, quantities, and types of materials managed. The system cannot integrate with company accounting systems, which can result in managing these costs separately, rather than integrating them with other company costs. The system is not designed for, and has no modules for, project management. Like most other waste management and tracking tools, the primary objective is to facilitate compliance, not to manage and minimize costs. However, if this area were developed, pollution prevention and other environmental initiatives would be encouraged as a result of better cost information.

Basis for evaluation: Based on telephone communication with Corbus and information package.

Contact information: Commercially available
Corbus
206 Line Road Kennett Square, PA 19348
Mary Smucker (Marketing Representative) 610-444-5691 610-444-6461 (FAX)

ABBREVIATED PROFILES

Software from Environmental Software and Systems Inc.

ES&S supplies a number of different software packages, including those that are repackaged EPA programs. These packages are for cost estimation, dispersion modeling, air emissions estimation and modeling, process analysis, hazard evaluations, etc.

CHEMMIX: Determines the effects of combining chemicals with

other chemicals, or with water. An unlimited number of chemicals can be analyzed, and the program comes with over 3 million reactions between 1700 chemicals. Additional chemicals can be added by users. This software can be used to alert users of a number of potential hazards, compatibility determinations, emergency response plans, contingency plans, spill response, and chemical shipments. The next update is expected in the fourth quarter of 1995.

The software is available in IBM (DOS, 384 KB RAM, 3 MB free disk space) and Macintosh (I MB RAM, 3 MB free disk space) formats. The cost of the software is $395.

CHEMICAL REACTION GUIDE: Aids in the identification of hazardous combinations.: $12.95.

EMERGENCY CALL SYSTEM: An IBM compatible hardware/software system designed to activate an emergency response network.; $2500.

PART B OUTLINE: Version 6.0 for DOS or Macintosh. Guidance and outline for writing and updating a Part B permit or permit application,: $49

RI/FS: For DOS or Macintosh. Guidance for submitting a RI/FS for RCRA sites. $49

ENVIRONMENTAL SOFTWARE LIBRARY (86-92): Relational database of all environmental software reviewed and summarized by Pollution Engineering since 1986. $30.00

ENVIRONMENTAL SOFTWARE LIBRARY (1994): Relational database of all environmental software published in Pollution Engineering in 1993. $49.00

ENVIRONMENTAL SOFTWARE LIBRARY (1995): $69

Spreadsheet Templates (MS-DOS and Macintosh)

1. Method 5 Stack Testing: Calculates emissions from a source by using data collected in a stack test.
2. Storage Tank: Calculates hydrocarbon emissions from fixed roof, internal floating roof, and external floating roof storage tanks.
3. VOC Calculations: Calculates actual coating line emissions in the RACT (Reasonable Available Control Technology) format, and in the NSPS (New Source Performance Standard) format.

All templates are priced at $39 each, or are available as a package at a special rate of $100.

Programs Developed for/by the USEPA (MS-DOS only)

1. CAT (Controlling Air Toxics): Equipment evaluation and selection for air toxics. Includes information on some equipment types and calculates design parameters and costs for each.
2. ISC2 (Industrial Source Complexes Version 2): One of EPA's accepted dispersion modeling methodologies for assessing pollutant concentrations from various sources associated with an industrial source complex.
3. SIMS (Surface Impoundment Modeling System): Estimates air emissions from surface impoundments.
4. LAND7: Analytical procedure for estimating compound-specific emissions from land treatment of organic compounds.
5. TANKS: Estimates emissions of organic chemicals from storage tanks according to EPA's AP-42.
6. SWAMI (Strategic Waste Minimization Initiative): Uses process analysis for identifying waste minimization opportunities in an industrial setting.
7. SCREEN: Air pollution dispersion model for applying screening

techniques to estimate the air quality impact of stationary sources.
8. HAP-PRO: (Hazardous Air Pollutant Program): Calculates capital and annual costs for up to 6 VOCs and three particulate control devices.
9. CHEMDAT7: Spreadsheet template that includes analytical models for estimating VOC emissions from TSDF processes.
10. POTW Expert: Advisory system for improving the performance of wastewater treatment facilities.
11. LANDFILL AIR EMISSIONS: method for approximating nonmethane organic compounds and individual air toxic gas emissions from landfills.
12. MECHANICAL DUST SOURCES: Calculates emission estimates from paved and unpaved roads, materials handling, agricultural tilling, and construction/demolition. Includes WIND EROSION program.
13. AIR TOXIC EMISSION FACTOR DB: Compilation of emission factors for toxic air pollutants and sources. Each costs $50 individually, or $400 as a package.

Other Software

1. LEAD (Uptake/biokinetic model for lead): Determines total lead uptake in children (0-6 years) from various sources.
2. ARCHIE (Automated Resource for Chemical Hazard Incident Evaluation): Assessment procedure for mitigating hazardous material accidents.
3. GEOPHYSICS ADVISOR EXPERT SYSTEM: A system for the non-geophysicist for applying geophysics to hazardous waste site investigations.
4. WW Process: wastewater treatment process programs for calculating different variables including settleable solids, waste gallons, BOD, etc.
5. METHODS: A database of all EPA, NIOSH, OSHA, ASTM, and other methods for sampling and analysis of toxics in ambient and

indoor air.
6. ORGANIC FATE IN WATER: Steady state and dynamic models for rapid evaluation of behavior of organic chemicals in lakes, rivers, and estuaries.
7. LC50: Estimates LC50 using the Trimmed Spearman-Karber method.
8. POLLUTANT FATE IN WATER: Set of routines that computes the direct photolysis rates and half lives of pollutants in the aquatic environment.
9. ACCESS EPA: A guide to EPA information resources, services, and products. 10. INDOOR AIR QUALITY: Estimates indoor air quality in a multiroom building.
11. TOXST (Toxic Modeling System Short Term): Addresses the problem of estimating expected exceedances in short-term health effects thresholds in the vicinity of continuous and intermittent toxic releases from a complex industrial site.
12. TOXLT (Toxic Modeling System Long Term): Evaluation of lifetime cancer risks and chronic noncancer hazards that may result from long-term exposure to toxic air pollutants.
All these programs cost $50 each.

Contact information

Environmental Software and Systems Inc. PO Box 1182
Bowling Green, OH 43402-1182 Roger Worth 419-353-8540 419-353-8540 (FAX)
Basis for information: Based on information received from supplier in May, 1995.

CHAPTER 8. FINANCIAL ANALYSIS SYSTEMS

This chapter organizes software reviews into four categories : Costing and Financial Analysis of Pollution Prevention Projects, Economic Analysis for Pollution Prevention, Pollution Prevention and Financing, and Pollution Prevention Benefits.

Costing and Financial Analysis of Pollution Prevention Projects

Prepared by: The Northeast Waste Management Officials' Association (NEWMOA) and the Massachusetts Office of Technical Assistance

Prepared for: Prepared under grant from the North-East Hazardous Substances Research Center (EPA funded)

Application: financial analysis: · Environmental impact analysis: - Waste management/P2: * Environmental cost listing/database: - Cost estimation: - Alternative product/process comparison: -

This method provides the format for the financial evaluation of pollution prevention options. The training manual in the updated version has industry-related examples.

Development/publication date and updates: 1992. Updated version, "Improving your competitive position," published in 1994, and comes in two styles - a Training Manual and an Instructors Guide.

Public availability: Available

Purpose and current use: NEWMOA has sold the manual and has also used it in training sessions organized through contract with

Waste Reduction Institute for Training and Applications (WRITAR). NEWMOA has used the manual in training workshops in various EPA regions around the country. It has also been used in some industry conferences.

Cost information: Old manual: For-profit organizations \$40 Non-profit organizations \$20 Training manual: \$15 Instructors Guide: \$30.

Contents and system requirements: Spreadsheet formats, case studies

Summary of methodology

The method consists of both a quantitative and qualitative analysis of the pollution prevention project. Quantitatively, costs are lumped into three major categories:

1. Direct material,
2. Direct labor, and
3. Overhead

Overheads include indirect materials and labor, including hidden environmental costs. Qualitatively, the method considers:

1. Intangible factors
2. Potential liability

Intangible factors include market share, employee/union relations, and shareholder relations, corporate image, community image). Potential liability includes disposal, storage, transportation, real property damage, civil actions, toxic tort suits, economic loss from remediation, and fines/penalties.

Life-cycle stages covered

Raw materials acquisition
Manufacturing
Use/reuse/maintenance
Recycle/waste management

Type of costs considered

Conventional
Potentially hidden
Contingent
External

The method looks at conventional and hidden costs of inputs and outputs, and liability costs of outputs (qualitatively). The updated manual also discusses the assessment of potentially hidden less tangible factors such as public image, shareholder relations, etc. These have also been defined as potentially hidden costs. It does not include external, that is, social costs into the analysis.

Method of cost estimation

The evaluation involves six steps:

1. Gathering complete, relevant cost information from the current process
2. Estimating relevant cost information for the proposed pollution prevention projects
3. Listing all differential costs resulting from the options
4. Applying financial analysis tool(s)
5. Interpreting the quantitative result
6. Considering the qualitative factors

The method calculates the total annual amount, represented as: (wage rate) x (number of hours spent for the activity) x (number of occurrences of the activity per year)

Generation of financial indicators

Net present value (NPV)
Payback period
Internal rate of return (IRR)
Benefits cost ratio
Others

The updated manual describes all these indicators along with methods of calculation. It also describes the accounting rate of return as an indicator. Capital budgeting, cash flow, and the economic lifetime of the project are considered in cost calculations. Discounted cash flow methods are used.

Attributes

The method performs qualitative analyses of activities not accounted for by conventional costs. The updated manual has been published in two parts. The manual itself contains basic course material. The Instructors Guide additionally contains teaching notes, exercises, overheads, and more case studies.

Limitations

Social (external) costs are not taken into account. The liability and potentially hidden less tangible costs (e.g., public image, etc.) that are taken into account are not quantified. The method does not assist in life-cycle costing.

Basis for evaluation: The information provided here is based on a profile of the tool presented in Weitz et al. (1994). Subsequent reviews were done by incorporating updated information from reports and through telephone communications with NEWMOA in May, 1995.

Contact information: Available from NEWMOA.
Tern Goldberg (Pollution Prevention Program Manager) NEWMOA 129 Portland Street, Suite 502 Boston, MA 02114-2014
617-367-8939 617-367-8558

The Massachusetts Office of Technical Assistance Executive Office of Environmental Affairs 100 Cambridge Street, Suite 2109 Boston, MA 02202 617-727-3260

Economic Analysis for Pollution Prevention

Prepared by: Pollution Prevention Division Washington State Department of Ecology

Prepared for: Washington State Department of Ecology

Application: financial analysis: · Environmental impact analysis: - Waste management/P2: - Environmental cost listing/database: * Cost estimation: - Alternative product/process comparison: -

This method compares pollution prevention options with current operating practices.

Development/publication date and updates: July 1992. A fact sheet on the same topic was published in 1995.

Public availability: A 4-page fact sheet is available

Purpose and current use: Information fact-sheet.

Cost information: The fact sheet is available free of charge through the Washington State Department of Ecology.

Contents and system requirements: Cost analysis spreadsheet formats, activities for analysis

Summary of methodology

The method recommends including only those differential costs that would change if the pollution prevention opportunity was implemented. The method also recommends a scoring system risk analysis with considerations based on risks such as property contamination, compliance violations, health effects, accidents, and long-term disposal liability. The updated fact sheet does not discuss this. "Total cost" is calculated as the usual cost of running an industrial process plus the environmental costs such as compliance and oversight or management. The updated version of the fact sheet requires users to enter estimates of compliance and oversight costs.

Life-cycle stages covered

Raw materials acquisition
Manufacturing
Use/reuse/maintenance
Recycle/waste management

The method considers raw materials acquisition, processing and manufacturing, transportation/storage/packaging, and waste management. The raw material acquisition stage is not check-marked, as our definition of this stage is different from the implication in this case. This is covered to the extent that product/vendor research is done. The updated version does not include waste management costs in the sample spreadsheet.

Type of costs considered

Conventional
Potentially hidden
Contingent
External

The method estimates conventional and hidden costs of inputs and outputs, and liability costs of outputs. Typical costs and activities are listed as follows:

Usual Costs

Depreciable Capital Costs engineering procurement equipment materials utility connections site preparation facilities installation

Operating Expenses start-up training -initial raw materials working capital raw materials supplies direct labor utilities, maintenance salvage value

Operating Revenues revenues by-product revenues

Compliance Costs

Receiving Area spill response equipment emergency response plan

Raw Materials Storage storage facilities secondary containment right-to-know training reporting and record keeping safety training container labels

Process Area safety equipment right-to-know training waste collection equipment emission control equipment sampling and testing reporting and record keeping

Solid and Hazardous Waste sampling and testing containers labels and labeling storage areas transportation fees disposal fees

Air & Water Emissions Control permit preparation permit fees capital costs operating expenses recovered materials inspection and monitoring record-keeping and reporting sampling and testing emergency planning discharge fees

Oversight Costs

Purchasing
product/vendor research regulatory impact analysis inventory control

Engineering
hazard analysis sampling and testing

Production
employee training emergency planning medical monitoring re-work waste collection disposal management inspections and audits

Marketing
public relations

Management
regulatory research legal fees
information systems penalties and fines insurance

Finance
credit costs tied-up capital

Method of cost estimation

Users estimated costs for activities. The following methods of estimation are recommended in the 1992 fact sheet:

1. Differential accounting: only the future cash flows that would change if the pollution prevention project was implemented are accounted for.
2. Inclusion of labor hours.
3. Extended time frame: a 5-year period is recommended for analysis.
4. Modified financial criteria: a lower rate of return, lower discount rate, or shorter payback period is recommended because of intangible benefits such as reduced risk, improved quality, fewer sick days, or better public image.

Generation of financial indicators

Net present value (NPV)
Payback period
Internal rate of return (IRR)
Benefits cost ratio
Others

Instructions and tables on how to calculate the NPV are provided.

Attributes

This method not only accounts for conventional costs, but also liability costs. Reduced risks are accounted for with a weighting system. The method in the 1992 fact sheet identifies risk factors, estimates their importance using a "risk score," and provides a way of using the results for decisions. High risks are weighted most heavily; unknown risks are given a "medium" score.

Limitations

External (social) costs are not included in the analysis. The sample spreadsheet calculates the NPV for a 5 year project. This time frame may be too small for some environmental projects.

Basis for evaluation: The information provided here is based on a profile of the tool presented Weitz et al. (1994). Subsequent reviews were done by incorporating updated product information obtained from the Washington State Department of Ecology in May, 1995.

Contact information; A 4-page fact sheet called "Cost Analysis for Pollution Prevention" is an update of the original paper and is available through the Washington State Department of Ecology. Ecology report - Publication # 95-400.
State of Washington, Department of Ecology PO Box 47775
Olympia, WA 98504
General number: 360-407-6730
Michael Johnson 360-407-6338

P2/FINANCE: Pollution Prevention Financial Analysis and Cost Evaluation System

Prepared by: Tellus Institute **Prepared for:** State of New Jersey

Application: financial analysis: · Environmental impact analysis: - Waste management/P2: - Environmental cost listing/database: * Cost estimation: - Alternative product/process comparison: »*

Analysis for the financial evaluation of pollution prevention (and other) projects.

Development/publication date and updates: 1993. Version 2.1 of P2/FINANCE was released in 1995.

Public availability: Available

Purpose and current use: This system was prepared as part of a project under the State of New Jersey. P2/FINANCE currently has 870 official copies' out of which 532 are businesses' (including consultants), 165 are in-state agencies,* and 165 are federal facilities.

Cost information: For government agencies (federal/state/local), the tool is available free of charge through the Pollution Prevention Information Clearinghouse. Tellus has license agreements with the New Jersey Department of Environmental Protection, Maine Department of Environmental Protection, and Massachusetts Toxics Use Reduction Institute from which businesses within the state can obtain the software directly for nominal charges. All other organizations that do not meet the criteria of the EPA or state agency license agreements can obtain the software directly from the Tellus Institute for the following charges:

National agencies: $ 10,000 State/regional agencies: $5000 Private firms:
>$100 million in worldwide sales: $1000 $10 - $100 million: $500 <$10 million: $200 Industry trade associations: $500 Non-profit organizations: $ 100 Individuals: ISO

Contents and system requirements: Software, user's manual.

- Windows Version 4.0
- IBM-compatible PC, XT, or AT
- MS Excel 4.0 for Windows Version 3.4a for DOS IBM-compatible

PC, XT, or AT
· Lotus 123 Version 3.1 or higher

Other requirements:

· Color monitor recommended
· Graphics printer, HP Laser Jet

These include copies that have licensed directly from Tellus, and those that have been obtained from an intermediate organization (such as the EPA) that has licensed the software tar distribution.

Businesses include those who have received the software from state agencies and Tellus. "In-state agency" refers to any state government agency who orders the software, e.g., the Department of Public Health.

Summary of methodology

The approach has four key features:

· a broad inventory of costs and savings
· the allocation of all costs and savings to specific process and product lines
· extended time horizons for the capture of long-term benefits
· the use of profitability indicators that incorporate the time value of money

The inventory of costs and savings helps users in building their analysis and including all costs and revenues.

Life-cycle stages covered

Raw materials acquisition

Manufacturing
Use/reuse/maintenance
Recycle/waste management

Type of costs considered

Conventional
Potentially hidden
Contingent
External

The methodology is primarily used to account for conventional and hidden costs of inputs and outputs. Liability costs can also be included by specifying the year in which the liability is expected to occur. Potentially hidden less tangible costs (e.g., public image) could be entered as part of the 'other costs' category. However, external (social) costs are not included in the analysis (In our definition, less-tangible (internal) costs are included as hidden costs). The following indicate the major cost elements that P2/FINANCE typically includes.

Capital Costs
Purchased equipment Materials
Utility Connections and new utility systems Site preparation Construction/installation Engineering/contractor Start-up/training Contingency Permitting
Initial charge for catalysts and chemicals Working capital Salvage value
Operating Costs
Direct materials
Waste management (materials and labor) Utilities Direct labor Other Regulatory compliance Insurance Revenues Future liability

Most of the costs listed above will include several sub-categories, with examples provided of additional cost elements that users can consider. For example, examples of sub-elements for revenue include change in sales from market share and corporate image. Social costs (externalities) are not included.

Method of cost estimation

The method provides blank templates for data input by users for current and alternative practices. Pre-defined category names are included. Operating costs are entered for both the current and alternative processes, while capital costs are entered for the alternative process only. Other optional inputs include debt and equity parameters, interest and tax rate estimates, and an escalation rate.

Generation of financial indicators

Net present value (NPV)
Payback period
Internal rate of return (IRR)
Benefits cost ratio
Others

Several financial indicators are calculated including NPV, IRR, and simple payback. The time horizons illustrated for each indicator are 5,10, and 15 years, as well as one user-defined time horizon.

Attributes

P2/FINANCE provides a mechanism to compare current and alternative practices. Its use is not limited to the analysis of P2 projects. The software comes equipped with an embedded cost list to remind users of potential costs and revenues that they might otherwise omit. Users can change field names in P2/FINANCE to tailor it to their requirements or facilitate the analysis of other projects. The familiar spreadsheet format facilities use by non-financial persons, such as process engineers. Demonstration disks of P2/FINANCE are available through Tellus. Tellus has a user-support technical hotline that is available free of charge to users.

Tellus is currently developing some industry-specific versions of PZ/FINANCE. These versions are different from the original spreadsheet version in that they are programmed in FoxPro, a database language, which makes the software much more flexible. Some of the new features:

· capital costs can be input for the current scenario
· profitability of multiple scenarios can be assessed
· capital costs can be input in any year
· variable operating costs can be defined
· cost items can be added to the predefined lists

These versions are based on Total Cost Assessment principles (such as version 2.1 of the spreadsheet), but contain process-specific tailored lists for the particular industries. For example, in the screen printer's version, P2/FINANCE for Screen Printers, Version 1.0, costs are entered in one of the major screen printing processes and the cost list contains relevant screen printing materials under the heading materials, such as ink. This specificity can better assist screen printers in estimating the profitability of pollution prevention projects.

P2/FINANCE for Screen Printers will be unveiled soon. It can be obtained through the Screen printers' trade association. Screen Printing and Graphic Imaging Association, International (SGIA). Soon Tellus will be working on three other industry-specific versions with funding from NIST: lithographic printing, printed wiring board, and metals fabrication and finishing.

Limitations

Users cannot enter any additional rows and columns to the spreadsheet. The system does not currently facilitate building-in of any algorithms. External (social) costs are not included in the analysis.

Basis for evaluation: The information provided here is based on a profile of the tool presented in Weitz et al. (1994). Subsequent reviews were done by incorporating updated product information obtained through communications with the Tellus lnst. in May 1995.

Contact information: Tellus Institute 11 Arlington Street Boston, MA 02116-3411
Angola Dierks Email: adierks@tellus.com Deborah Savage, Ph.D Email: dsavage@tellus.com 617-266-5400 617-266-8303 (FAX)
P2/Finance is available free of charge for federal, state, and local government agencies through the Pollution Prevention Information Clearinghouse: 202-260-1023

Pollution Prevention Benefits Manual

Prepared by: ICF Incorporated

Prepared for: Office of Solid Waste/Office of Policy Planning and Evaluation, U.S. Environmental Protection Agency

Application: Financial analysis: · Environmental impact analysis: - Waste management/P2: * Environmental cost listing/database: * Cost estimation: - Alternative product/process comparison: -

The manual was designed to help industry incorporate environmental

costs into decision-making. The objective was to help industry increase environmental protection and reduce compliance costs by using true costs of current materials and waste management and by evaluating financial paybacks of pollution prevention alternatives.
Development/publication date and updates: October 1990

Public availability: Available

Purpose and current use: The manual was produced under the direction of Dr. Ron McHugh, Office of Policy Planning and Evaluation, U.S. EPA by ICF Incorporated. The manual is being distributed through the Pollution Prevention Information Clearinghouse. A recent study showed that out of 50 companies estimating environmental liabilities, 20% were using the P2 Benefits Manual method. The rest were using methods created inhouse.

Cost information: Free from the Pollution Prevention Information Clearinghouse (PPIC).

Contents and system requirements: Consists of a manual and worksheets.

Summary of methodology

This method compares pollution prevention alternatives as compared to the current practice. The method sets up a hierarchy of costs. The costs in the hierarchy progress from obvious internal to not-so-obvious hidden costs. The manual stresses that companies need not address all the tiers in every analysis. Rather, companies should include just enough tiers to show the economic gain from choosing the P2 product or process. At each tier of the hierarchy, users complete worksheets that identify inputs and outputs to be estimated and provide formulas with suggested values based on user-specified situations.

The hierarchy of costs is as follows:
TIER- Tier 0-Usual Tier I -Hidden Tier 2-Liability Tier 3-Less Tangible

EXAMPLE: equipment, labor, materials compliance, permitting penalties/fines, liabilities consumer response, employee relations
Although the costs in Tier 3 have been referred to as less tangible in the manual, we include them as hidden costs (since they are internal). However, in the calculation of liability costs events such as natural resource damage is taken into account. This can be classified as a social or external cost. Information taken from the document "Environmental Cost Accounting for Capital Budgeting: A Benchmark Survey of Management Accountants," 1995, prepared by the Tellus Institute for EPA, OPPT.

Life-cycle stages covered

Raw materials acquisition
Manufacturing
Use/reuse/maintenance
Recycle/waste management

The manual considers on-site waste generation and waste management on- and off-site. The second stage (manufacturing) is considered to calculate possible future liability costs, according to the manner of disposal and applicable regulations. Use/reuse does not include the use of a product once it leaves the company.

Type of costs considered

Conventional
Potentially hidden

Contingent
External

The manual considers conventional and hidden costs of inputs and outputs, liability costs of outputs, and the impact costs of consumer response, employee relations, etc.

Method of cost estimation

Costs in Tiers 0 and I are estimated users. The model provides cost equations to assist in calculating future liability costs. Each equation consists of several variables for which the manual provides suggested values. Liability equations incorporate user-identified values and are provided for the following:

· soil and waste removal and treatment,
· ground-water removal and treatment,
· surface sealing,
· personal injury,
· economic loss,
· real property damage, and
· natural resource damage.

To calculate less-tangible costs, the manual recommends users use their own judgments, and provides methods of making percent adjustments to expenses and operating revenues. However, users must decide the number of tiers they want to include in their analysis.

Generation of financial indicators

Net present value (NPV)
Payback period
Internal rate of return (IRR)

Benefits cost ratio
Others

The report guides users in preparing incremental annualized cash flows and calculating after-tax total annualized savings, NPV, and the IRR. It suggests generating these figures after calculating each tier of costs, which can be very useful for users. The worksheet allows users to calculate NPVs using four different discount rates, one of which is user-specified.

Attributes

The model considers that consumer acceptance and employee/union relations change when wastes are reduced. Further, the tiered approach provides a progression of detail for users that may aid in comparing previous costing methods. The appendix provides an example case of a hypothetical firm. The section on hidden regulatory costs is discussed in detail with brief descriptions of specific requirements in the regulatory cost protocols of six acts (RCRA, CERCLA, Superfund, CM, CWA, OSHA).

Although the manual does not go into detail on the Tier 0 facility analysis and cost development, it refers readers to another manual prepared by EPA, *Waste Minimization Opportunity Assessment* Manual (EPA/625/7-88/003), that covers the subject in detail.

Limitations

The manual discusses only certain aspects of calculating liability costs. The assumptions behind formulating the given equations and references for additional information are not provided. Users are expected to use assumptions and equations that suit their particular cases. However, the manual makes an important point that liability cost calculations should not be left out only because they are difficult to calculate. The section on hidden regulatory costs under the six acts is now outdated and will need to be updated by users to reflect changes in regulations since 1969. Calculations can become cumbersome without a software system. However, some simplifying tables are provided. The manual does not discuss the pros and cons of using one financial indicator as a decisive factor over the other.

Basis for evaluation: The information provided here is based on a profile of the tool presented in Weitz et al. (1994). Subsequent reviews were done by incorporating updated product information from a copy of the manual (Volume 1, October 1989), and through telephone communications with the U.S. EPA in April, 1995.

Contact information: Holly Elwood USEPA, Office of Pollution Prevention and Toxics 401 M. Street, SW Washington, DC 20460 202-260-4362
Report number: EPA230-R-89- 100 Pollution Prevention Information Clearinghouse 202-260-1023
Joseph Karam/James Faulstich/James Dickson ICF Technology, Inc. Fairfax, Virginia 22031-1207 703-934-3000

PRECOSIS

Prepared by: George Beetle, George Beetle Company

Prepared for: U.S. Environmental Protection Agency, Center for Environmental Research Information, Cincinnati, Ohio

Application: Financial analysis: · Environmental impact analysis: - Waste management/P2: - Environmental cost listing/database: - Cost estimation: - Alternative product/process comparison: *
Financial analysis of waste reduction projects.

Development/publication date and updates: 1989. There have been some enhancements and updates.

Public availability: Available

Purpose and current use: The software was designed to support participants at a series of seminars on waste minimization sponsored by the EPA in 1989. Approximately 1000 copies of the older version (uncompiled and more difficult to use than the current version) were distributed at the seminars. The method has since been enhanced and placed in general distribution under license by the George Beetle Company.

Cost information: $50

Contents and system requirements: Manual and software. A menu-driven program consisting of ten data input tables and two output reports containing financial calculations. Requires an IBM-compatible computer with 512 KB (or more) of base memory.

Summary of methodology

The method first prompts users to describe the current process. Then they enter cost data for labor, materials, facilities and waste management, revenue data, and financial parameters.

Cost data are grouped into three categories: « resources effects (labor, materials, facilities)

· revenue or value effects (changes in output quantity and quality, secondary products and services)
· waste management effects (storage, handling, disposal, compliance, insurance, litigation)

Users need to enter unit cost data for labor, materials, facilities, and waste management. Also, users need to specify the number of units needed for the current process. Users are asked for cost information concerning the following activities for the current practices and the alternative practices:

1. labor resources and labor requirements by labor class, average hourly labor rates, benefits by labor class
2. materials usage rates, delivery costs
3. facilities equipment and building sizes and capacities, average service life, typical salvage value, usage rates
4. revenues from current production rates for primary and secondary products and expected changes for alternative practices, revenue rates for products
5. waste management quantities, waste management costs including monitoring and reporting, insurance for workers and third parties

PRECOSIS calculates various financial indicators to evaluate the profitability of alternative processes. Data on other alternative strategies can be entered to compare financial differences between alternatives (can compare up to 5).

Life-cycle stages covered

Raw materials acquisition
Manufacturing
Use/reuse/maintenance
Recycle/waste management

These life-cycle stages are considered within the confines of a company.

Type of costs considered

Conventional
Potentially hidden
Contingent
External

Method of cost estimation

Users enter cost data for labor, materials, facilities and waste management, revenue data, and financial parameters. To calculate costs associated with the alternative process, users need to enter an expected increase or decrease (called 'effects') in the number of units used, waste generated, and product produced as a result of the process change. The software calculates the net present value, estimated payback period, internal return on investment, and other financial indicators.

Generation of financial indicators

Net present value (NPV)
Payback period
Internal rate of return (IRR)
Benefits cost ratio
Others

Attributes

PRECOSIS facilitates the side-by-side comparison of up to five waste reduction alternatives. Improvements in production due to waste reduction alternatives are considered. Facility and equipment salvage values are also entered into the equations. The software is a menu-driven program consisting of ten data input tables and two output reports containing financial calculations.

Limitations

The approach does not consider contingent and external costs. It is not designed for life-cycle cost analysis. Users need to spend some time understanding the methodology and structure of PRECOSIS (different from conventional project financial analyses). Most of the calculations are not explained in the user's manual, nor would the typical user be able to examine the formulas in the software.

Basis for evaluation: The information provided here is based on a profile of the tool presented in Weitz et al. (1994) and in White et al. (1991). Subsequent reviews were done through telephone communications with George Beetle in May 1995.

Contact Information: *533* Arbutus Street Philadelphia, PA 19119 George Beetle 215-438-0598 215-438-7876 (FAX)

ABBREVIATED PROFILES

PAPA Investment Model

This is a multi-objective, mathematical programming model to maximize pollution prevention, cost and energy savings, and minimize investment and life-cycle costs. It is built around the Army / DOD environmental policy goals. The model was developed by the Army Concepts Analysis Agency (CM), in October, 1994.

The program develops and evaluates investment strategies in support of environmental policy and program requirements, using previously accumulated data and incorporating policy and budget parameters. Project cost benefit data is collected from all available sources such as hazardous waste minimization reports, databases, completed project economic analyses, etc. Where the cost benefit data was absent, it was estimated from comparable projects with assistance from the Army Environmental Center (AEC). Types of costs taken to account include procurement, facility modification, and labor costs, and recurring (operation and maintenance) costs. Annual pollution reduction (tons per year) can also be recorded. Annual savings or payback periods can be computed. The definition of life-cycle costs used here is the total (full) internal costs of a project. The model can incorporate changes in policy, budgetary, and technical data elements to develop and evaluate alternate investment strategies.

The program runs on an IBM workstation using an IBM optimization package. The program is primarily being used by the Army Chief of Staff for Installation Management. It has not as yet been adapted for use by other organizations or Federal facilities.

Contact Information: US Army Concepts Analysis Agency 8120 Woodmont Avenue Bethesda, MD 20814-2792
Jim Connelly - 301-295-1682 Maj. Tyrone Alien - 301-295-1443

FAX: 301-295-1662.

Basis for evaluation: Based on information on tools received from the DOD and telephone communications with Jim Connelly in August, 1995.

CHAPTER 9. ENVIRONMENTAL COSTING IMPACT ANALYSIS SYSTEMS

APACS Life Cycle Cost Software Demonstration

Prepared by: Moore Products Co.

Prepared for: Use as a marketing tool for their product-the Advanced Process Automation Control System (APACS).

Application: Financial analysis: · Environmental impact analysis: - Waste minimization/P2: - Environmental cost listing/database: - Cost estimation: - Alternative product/process comparison:

Calculating life-cycle costs (LCC) of APACS control system. The program can be used as an example of a method for financial analysis of product life-cycle costs for capital budgeting decisions. The program does not cover all life-cycle stages or all cost categories, but users can incorporate such information in their versions.

Development/publication date and updates: 1994. No updates

Public availability: Available

Purpose and current use: Not a commercial product. Designed to help users evaluate the APACS system.

Cost information: Available free of charge with APACS information brochure.

Contents and system requirements: The disk runs on a PC and requires MS Excel (version 4.0 or greater) for Windows.

Summary of methodology

This software product has been brought out by Moore Co. for educational purposes to allow users to conduct a life- cycle cost analysis on their computer control system and compare that with Moore's product, APACS. The system is not commercially available. The disk was brought out to demonstrate the principles discussed in an article "Just How Expensive is Your Control System," in the November 1994 issue of *INTECH* (volume 41, number 5, page 33).

The demonstration disk consists of three files. The first calculates life cycle costs and guides users through an exercise that already has default values (users can change these). The system prompts users for input information related to design, purchase, installation, start-up, and yearly costs. It calculates the present value of all the costs. The second file calculates the safety and reliability metrics of several common system architectures. The inputs include failure rates, diagnostic coverage factors, and common cause factors. The third file is a spreadsheet, similar to that in the first file, and includes a number of input variables grouped in categories and calculates LCC as a function of these.

Life-cycle stages covered

Raw material acquisition
Manufacturing stage
Use/reuse/maintenance
Recycle/waste management

The definition of a life cycle used here is the total cost of a process/product from the initial outlay through all related future costs (maintenance, replacement and other operations) that may be incurred as a consequence of the initial investment decision. Thus, it would actually cover only one stage in the life cycle, the use/reuse of a process control system, including the initial purchase cost. No effects (if any) external to the company are considered. Information on the efficiency and key features is not included. However, examples are provided of how safety and reliability metrics related to systems can be calculated, with some common systems as examples.

Type of costs considered

Conventional
Potentially hidden
Contingent
External

The methodology considers conventional costs and hidden costs associated with maintenance, failure, and operating expenses. Since the program also includes an Excel spreadsheet, within which users can include estimates for other cost categories. However, they would need to develop these estimates on their own. The same structure can also be used to compare two completely different items-products, materials, or projects. However, since Excel is not designed as an estimating program, users should enter project totals for simplicity.

Method of cost estimation

As discussed in the software summary, the system consists of Excel files that list input variables and default values. It uses unit costs and time, or data plugged into list estimates. The formulas used can be viewed and analyzed. Users can customize the files to include information that reflects their applications. The system does not help in general cost estimation, other than providing default values related to the APACS system.

Generation of financial indicators

Net present value (NPV)
Payback period
Internal rate of return (IRR)
Benefits cost ratio
Others

The files contain formulas to perform various arithmetic functions (totals and subtotals) and to calculate present values of all future expenses. The software does not compute NPV because no benefit information is entered. However, users can always modify the formulas and information to do so. Excel has financial functions that include PV, NPV and IRR.

Attributes

The program includes functions performed within an Excel spreadsheet. Even though the definition used for LCC is different from ours, the software gives a useful example of total cost calculation. The INTECH article points out that most users do not take total costs into account, despite the intuitive logic for doing so. This is due to the uncertainty, difficulties and unfamiliarity involved in estimating costs such as future maintenance and failure costs. This is probably true of other cases as well.

Limitations

The system does not discuss the issues involved in the choice of discount rate and the example given does not take inflation into account. No life-cycle environmental cost information or estimating methods are provided. The recycle/waste management stage is not included in the analysis. The system and the cost estimates provided and the discussion serve a narrow purpose: to demonstrate life-cycle cost calculation for process control systems. If users want to apply this for other applications, they will need to develop the required information. Since this is not a commercial product, there is no user-support for applications that are not related to the original purpose.

Basis for evaluation: Software disk and information package received in early 1995.

Contact information: Not a commercial product
Moore Products Co. Spring House, PA 19477-0900 215-646-7400
215-283-6358 (FAX)
Intech article reprint information: Instrument Society of America
P.0. Box 12277 RTP, NC 27709
919-990-9282 919-549-8288 (FAX)

ECO-SYS

Prepared by: Randan D. Watkins/Ted Wheelis

Prepared for: Sandia National Laboratories (SNL)

Application: financial analysis: - Environmental impact analysis: * Waste management/P2: - Environmental cost listing/database: - Cost estimation: - Alternative product/process comparison: *

SNL is involved in the development of an information and expert system called EcoSys that performs impact analyses of product design and manufacturing processes. This program will help process and product engineers assess the environmental consequences of their decisions. The software can provide decision support in a number of areas including comparative process and product assessments, process optimization, and environmental needs assessments.

Development/publication date and updates: Version I of EcoSys was released in October, 1993. Version 2 was released in 1994. Various papers are being written focusing on the EcoSys software development.

Public availability: Available

Purpose and current use: In March, 1983, SNL brought out a document called "IMDIECM Life-Cycle Costs Part 1. Environmental Impact Metrics Definition, Stakeholder Survey and Panel Review." This document was prepared as part of an SNL program. Integrated Manufacturing and Design Initiative (IMDI). In the study, the panel developed an IMDI Environmental Impacts Model that builds on earlier SETAC work. The panel discussed the possibility of using Colby's (1990) five environmental management paradigms as a basis for assigning weights to environmental impacts. This exercise has since led to the development ECO-SYS. The software is being used in some private companies, primarily with the objective of evaluating the software and providing feed-back to SNL.

Cost information: EcoSys is currently available free of charge. DOE may consider commercializing the software at a later date.

Contents and system requirements: The Eco-sys software operates on a Sun workstation, using X-Windows. Users can remotely access the system using their PCs, Macs, or other SPARCstations.

Summary of methodology

The environmental modeling in ECO-SYS employs principles of life-cycle analysis as defined by the Society of Environmental Toxicology and Chemistry (SETAC) and the USEPA. The approach is structured around the life-cycle inventory. The system includes environmental impact decision models that are based on the application of the Analytic Hierarchy Process. The goals of EcoSys are identified as the following.

· identify and quantify the relationships between product design, material consumption, and waste generation, within a manufacturing process.
· apply these relationships to compare environmental consequences of competing product designs and processes.

There are three databases in the system. The first contains **process knowledge,** which assists users in building a tree (product/process hierarchy) that incorporates process and material information. Alternate processes can be built-in from this point. Ideally, the bottom level of the hierarchy should contain the most basic constituent sets of processes and materials.

The second database contains information that links material information to environmental impacts. Thus, processes can be defined by their environmental impacts. An in-built materials library assigns values to each of the environmental attributes for a set of selected materials. This takes to account the quantities into a process, and quantities out of a process. Currently, the EcoSys library consists of over 400 single and multi-component materials classified according to a wide variety of organic and inorganic chemical families.

Finally, the third database contains information on three different impact models. The impact models are built around three of Colby's paradigms: Environmental protection. Resource management, and Ego-development. Users can compare the different conclusions through the application of these paradigms. The models are based on the application of the Analytic Hierarchy Process (AHP), which supports to the subjective nature of environmental impact analysis.

The criteria that were selected for inclusion in these models are based on:

· **Environmental risk;** issues include global warming, ozone depletion, land-use, species extinction, and air and water toxics.
· **Waste production:** issues include quantity, percentage hazardous, disposition route, and fugitive emissions.
· **Resource consumption:** issues include quantity, energy, renewable materials use, and reuse.

Sub-criteria under risk that were given the maximum priority were derived from the findings reported by an EPA Science Advisory Board study 9 on risks associated with global environmental problems. Weights were assigned based upon interpretations of the three different paradigms.

Life-cycle stages covered

Raw materials acquisition
Manufacturing
Use/reuse/maintenance
Recycle/waste management

EcoSys contains inventory information only for the manufacturing stage of the life-cycle. However, the environmental attributes were chosen in such a manner to permit the creation of a life-cycle perspective, without complete inventory data. Thus, full life-cycle inventories are considered only in a more qualitative sense.

Type of costs considered

Conventional
Potentially hidden
Contingent External

The "costs" associated with these impacts were not evaluated.

Method of cost estimation

The IMDI report discusses a weighting method of cost estimation based on Colby's five environmental management paradigms: frontier economics; environmental protection; resource management; eco-development: deep ecology. The ECO-SYS software is built around three of these paradigms: environmental protection, resource management, and eco-development. However, no costs are estimated in ECO-SYS. Associated with each paradigm are differing philosophies of human nature relationships. The paradigms are overlapping and encompass several schools of thought. The following is a description of the possible environmental costing methodologies under each of the three paradigms considered in ECO-SYS.

Environmental Protection

· Taxpayers (public at large) pay environmental costs.

· Analytic modeling and planning methodologies include environmental impact assessment after design, optimum pollution levels, equation of WTP and compensation principles. EPA/SAB, 1990. Science Advisory Board, "Reducing Risk: Setting Priorities and Strategies for Environmental Protection." U.S. EPA, Washington DC, EPA SABEC 90-021.

· Economic analysis is based on the neoclassical model of the closed economic system. Ecological benefits are difficult to quantify, so environmental management in this paradigm is treated strictly as an added cost.

Resource Management

· "Polluter" (producers and consumers) pay environmental costs.

· Analytic modeling and planning methodologies include natural capital, true (Hicksian) income, maximization of United Nations System of National Accounts, ecosystem and social health monitoring, and linkages between population, poverty, and environment.

· Economic analysis is based on an extension of neoclassical economics that incorporates all types of capital and resources-biophysical, human, infrastructural, and monetary-into calculations of national accounts, productivity, and policies for development and investment planning.

· Pollution can be considered a "negative resource" (causing natural capital degradation), rather than as an externality.

· The concern for nature stems from the fact that hurting nature is beginning to hurt economic man. Environmental expenditures are considered necessary to avoid "more" costs.

Ego-Development

· A "pollution prevention pays" concept rewards those who do not pollute. The economy is structured to reduce pollution as a throughput.

· Analytic modeling and planning methodologies include ecological economics; open-system dynamics; socio-technical and ecosystem process design; integration of social, economic, and ecological criteria for technology; and trade and capital flow based on community goals and management.

· The relationship between society and nature can be considered a "positive sum game." Human activities are organized to be synergistic with ecosystem processes and services.

· Emphasis is placed on efficient, clean, renewable energy sources, increasing environmental information, community consciousness, and experiential quality of economic activity.

· An example of the eco-development paradigm is the International Joint Commission (IJC) between U.S. and Canada, which explicitly uses a stakeholder, positive-sum approach.

Generation of financial indicators

Net present value (NPV)
Payback period
Internal rate of return (IRR)
Benefits cost ratio
Others

Attributes

The EcoSys software provides a method by which impacts of different manufacturing processes and materials can actually be assessed and compared under three different paradigms. Thus, the assessment methodology goes beyond the quantification of input and output streams. EcoSys focuses on the process for performing impact analyses, the reasoning being that the impacts result from human activity (processing) rather than from the materials or products themselves.

Limitations

No impact costs are estimated. The software uses a rating system that typically assigns values of I for low impact, 5 for moderate impact, and 9 for high impact. This will be refined as more data on the environmental impacts becomes available. EcoSys currently contains inventory data only on the manufacturing stage of the life-cycle. As customer- supplier relationships evolve, Eco-Sys can better support detailed analyses of upstream and downstream effects. The Eco-sys software is still in the developmental stage.

Basis for evaluation: The information provided here is based on information received from SNL in May, 1995.

Contact information: The report and software are available through SNL. A demonstration version of the software is also available.

Sandia National Laboratories Org. 6611 Albuquerque, NM 87195-5800
Randan Watkins Ted Wheelis 505-845-9298 505-844-0116 (FAX)

EE Energy/Environmental Life-Cycle Assessment

Prepared by: Pacific Northwest Laboratory, and National Renewable Energy Laboratory

Prepared for: U.S. Department of Energy, Office of Energy Efficiency (EE) and Renewable Energy

Application: Financial analysis: - Environmental impact analysis: w* Waste minimization/P2: - Environmental cost listing/database: - Cost estimation: - Alternative product/process comparison:

Total Energy Cycle Assessments (TECAs): A specialized form of energy/environmental life-cycle assessment that facilitates the evaluation of the supply and demand sides of energy systems. This handbook addresses the management, methodology, and tools being developed by EE to support TECAs. Discusses basic approaches for evaluating life-cycle environmental impacts associated with technologies. Applications can extend to non-DOE projects as well.

Development/publication date and updates: July 1993. Several other follow-up studies have been done, including the development of computer models for LCA.

Public availability: Available

Purpose and current use: Developed as part of the DOE'S on-going efforts to respond to the National Energy Policy Act of 1992 which required that energy decisions should include costs of health and ecological impacts. Public distribution of the report was not an objective. The primary purpose was to present LCA methodologies for potential use by the EE office.

Cost information: Available free of charge, in a limited manner.

Contents and system requirements: Worksheets, guidelines

Summary of methodology

A TECA examines the environmental impacts of energy systems, considering various life-cycle stages. The near- term focus of TECAs is the accounting of inputs and outputs of material, energy, and environmental residuals associated with each unit of "use" service.

The method consists of first conducting an environmental inventory analysis. TECA worksheets are provided to aid in identifying the following inputs and outputs.

· Inputs: energy, labor, and materials
· Outputs: energy, material and residuals

Data quality is assessed through indicators and a recommended matrix format. The inventory items are then translated into an inventory of impacts. Two steps are involved in the estimation of impacts:

1) the estimation of changes in ambient environmental conditions and
2) the estimation of health, ecological, and other impacts.

Step 1 predicts the level of exposures that will be realized for the specific environmental conditions being considered. Step 2 is used in conjunction with Step I and dose-response relationships to determine the associated impacts.

Finally, an economic valuation of the impacts is performed. This is done by measuring the changes in well-being of those either "damaged" or "benefiting" from the environmental or health impacts.

Life-cycle stages covered

Raw materials acquisition
Manufacturing
Use/reuse/maintenance
Recycle/waste management

The following life-cycle information is considered:

· raw materials acquisition
· transportation (of materials)
· processing (conversion to a useful energy form, for example, electricity)
· transportation (of the energy form)
· use

Assumptions about the waste management and pollution control methods used can be factored into the analysis.

Type of costs considered

Conventional
Potentially hidden
Contingent
External

The methodology discusses various methods for estimating non-market goods, but does not actually perform the valuation in any examples. It is suggested that capital and O&M costs be included in the technology characterization database, as part of an integrated TEC tool.

Method of cost estimation

The economic valuation suggested is in terms of "willingness-to-pay" (WTP) for or "willingness-to-accept" the changes in well being. WTP is defined as the maximum amount of money a consumer would be willing to pay to avoid the change, before it happened. Travel cost techniques and hedonic techniques are also discussed.

Generation of financial indicators

Net present value (NPV)
Payback period
Internal rate of return (IRR)
Benefits cost ratio
Others

Attributes

The report discusses various energy and fuel cycles and other elements of TECAs. Various methodological and data issues are discussed. The method attempts to identify and value environmental impacts. Examples of objectives and environmental impact issues are discussed from EE's perspective.

Limitations

There are no case-study examples of non-market valuation. The document does not attempt to provide a complete description of all the methods used to estimate these costs. The focus is on environmental impact analysis rather than full cost accounting.

Basis for evaluation: The information provided here is based on a profile of the tool presented in Weitz et al. (1994).

Contact information: Pacific North-west Laboratory MSIN K8-15, 3230 0 Avenue Richland, WA-99352
Ken Humphreys (PNL) 509-372-4279 509-372-4378 (FAX)

The EPS Enviro-Accounting Method

Prepared by: Bengt Steen- Swedish Environmental Research Institute Sven-Olof Ryding - Swedish Federation of Industries

Prepared for: Swedish Waste Research Council (AFR)

Application: Financial analysis: - Environmental impact analysis: * Waste management/P2: - Environmental cost listing/database: - Cost estimation: - Alternative product/process comparison.

A method of valuing various types of environmental impacts from emissions and use of resources. This can be used to help determine less-tangible costs in projects and operations.

Publication/development date and updates: December 15,1992 (Version 2)

Public availability: Report available.

Purpose and current use: The methodology was developed to provide industry with a tool to develop products in a sustainable manner by allowing them to calculate the ecological consequences of alternative products and processes. It can also be used by policy-makers to put costs on environmental impacts. Information on current use is not available.

Cost information: Information not available

Contents and system requirements: Not software-based

Summary of methodology

The approach used is to describe impacts on the environment in terms of one or several safeguard subjects and value changes in them according to the willingness to pay to restore them to their normal status. The five safeguard subjects are

· human health,
· biodiversity,
· production,
· resources, and
· aesthetic values.

Activities are then valued according to their estimated contribution to the changes in the five safeguard subjects. The EPS Enviro-Accounting method

· sets a value to a change in the environment, and
· estimates the contribution that resource depletion, emissions, or other activities will have on the environment.

Life-cycle stages covered

Raw materials acquisition
Manufacturing
Use/reuse/maintenance
Recycle/waste management

Background information is derived from LCA-based inventories of the materials and processes under review. However, the methodology itself does not categorize impacts as they occur in various life-cycle stages of a process or product. Much of the EPS Enviro-Accounting method is applicable to life-cycle cost integration and to the development of the E-CYCLE.

Type of costs considered

Conventional
Potentially hidden
Contingent
External

The methodology can be used for valuing external intangible costs. Costs of environmental impacts on the five safeguard subjects listed previously are evaluated. The report includes estimated indices for a wide range of environmental impacts. The system is not meant to include internal company costs: it is designed to value external environmental impacts.

Method of cost estimation

WTP to avoid environmental effects is used to estimate costs. Impacts are valued on a relative scale using environmental load units (ELU) according to the WTP for avoiding negative effects on the safeguard subjects. Using life-cycle data, users can calculate environmental indices for the materials and processes being studied. Users can also include qualitative information for describing how the indices were estimated. The values are not absolute figures but rather points of reference for further analysis and refinement.

The output from the EPS system is a value, based on a common metric, for different environmental impacts. The value may be broken down into its individual components for further analysis and users can determine the level of detail desired. The authors address several criticisms of using monetary values to assess environmental impacts. These include the following:

· the large gap between rich and poor in terms of disposable resources for environmental care, · needs of today often outweigh the needs for tomorrow,
· insufficient knowledge to value environmental impacts because the full consequences of impacts are not fully understood, and
· monetary valuation focuses on the needs of humans.

Environmental impact valuation is described as a subjective matter that can be given some degree of objectivity by studying decisions made in society or by surveying people's opinions. Contingent valuation is cited as a method for generating a relative rating of various environmental effects. Contingent valuation is used here to determine individuals' VVTP to avoid certain environmental effects.

Generation of financial indicators

Net present value (NPV)
Payback period
Internal rate of return (IRR)
Benefits cost ratio
Others

Attributes

Some positive characteristics of the tool are as follows:

· The indices and databases can be easily supplemented and updated as knowledge of environmental issues increases and changes.
· Sensitivity and error analyses can be performed.
· The outcomes can be grouped in various levels of aggregation to facilitate comparisons, as per user requirements.

The EPS system goes a step beyond the complex flows in a life-cycle analysis to provide users with easy to understand and manageable information.

Limitations

The method itself does not evaluate various alternatives, rather it supports decision-making by providing users with information that they can further use in analysis or further refine to suit their requirements. It is aimed to be used to value current and foreseeable impacts within legal limits (not sudden spills or accidents). The choice of unit effects also has certain limitations, such as the ability for the effects to be quantitatively valued. Additionally, the authors acknowledge the difficulty in assessing the actual value of an ecosystem in its entirety. The EPS method does not generate financial indicators such as NPVs, because it assesses environmental impacts alone-not costs and benefits.

Basis for evaluation: "The EPS Enviro-Accounting Method. An application of environmental accounting principles for evaluation and valuation of environmental impact in product design." Bengt Steen, Sven-Olof Ryding. Swedish Environmental Research Institute (IVL).

Contact information: AFR-REPORT nr 11. ISSN 1102-6944. ISRN AFR-R-11-SE
Swedish Environmental Research Institute (IVL) P.0 Box 47086 Gagjamningsgaten I S-402 58, Goteborg Sweden
Bengt Steen +46-031-46 00 80 +46-031-48 21 80 (FAX)

Hazardous Material Life-cycle Cost Estimator (HAZMAT)

Prepared by: The Analytic Sciences Corporation (TASC)

Prepared for: U.S. Air Force-HCS/EMP. (Human Systems Center,

Brooks Air Force Base)

Application: Financial analysis: - Environmental impact analysis: - Waste management/P2: * Environmental cost listing/database: - Cost estimation: Alternative product/process comparison.

An analytical tool to support trade-off analyses in selecting alternatives to hazardous materials.

Development/publication date and updates: March 1991. Version 3.0 released in December, 1993.

Public availability: Available to Federal facilities

Purpose and current use: The purpose of the system is to facilitate cost trade-off analyses for comparisons between one hazardous substance and another hazardous/nonhazardous substance. The objective is to reduce the kinds and amounts of hazardous materials used in weapon systems and their associated development, production, maintenance, and decommissioning processes. HAZMAT is manufacturing and maintenance process oriented.

Cost information: Free of charge.

Contents and system requirements: HAZMAT is programmed in Ada using the Ada-Sage programming tool and compiler.

Minimum system requirements are:
· IBM-compatible computer (386 recommended) MS-DOS 3.1 (or higher) 640K RAM
· 20MB hard drive (HAZMAT requires - 5MB of storage space)
· EGA/VGA color monitor

Summary of methodology

Users enter chemical-specific information, after which the system allows them to perform trade-off studies of one hazardous material versus another. The goal of this method is to include hazardous materials cost elements in four phases of the system: development, production, operation and support, and system disposal. Cost elements are identified for each phase for the inputs and outputs of different life-cycle stages. Model outputs are either documentation or cost outputs. HAZMAT functions in the following manner

1. Users enter information in the HAZMAT **main menu.** Options include generating a new estimate, modifying a previous on, viewing and/or print reports, entering a user inflation table, or exiting HAZMAT.
2. Users enter **general estimate information** such as the title, date, base and cost input years, discount rate, inflation table selection, and type of analysis. The analysis types are either modeling the cost of hazardous materials for one subsystem 10 for one year, or total phase 11 cost calculation.
3. Users select the **systems, types, subsystems and phase.** The phase choices are acquisition, operating and support, and decommissioning. The system moves on to different screens depending on the phase choice. *Note : A system is the major classification, such as an aircraft. The type is the major subdivision, such as a fighter plane, and the subsystem is the major subdivision of the type (e.g., Airframe). A phase represents the period of interest in the life of the subsystem.*
4. The system enters into a process loop in which users select one or more processes of interest, either from the HAZMAT database or by adding or modifying processes. Processes include items such as depot painting, repairing, base aerospace ground equipment, etc. In the process information screens, users can either use default values from the database or modify the information.
5. HAZMAT then moves into a **substance** (the materials-Ce.g.,

lubricating oil, cleaning compound, etc.] loop, where substances can be arranged into up to three groups with up to five substances each (baseline and alternatives). Again, users can choose substances from the database or add or modify substances. Entering the custom substance would involve entering related data such as quantity used in phase or process, disposal percentages, etc.

The reports generated include parameter list reports (general, process, and substance information), a base-year dollars report (reflect the purchasing power for that year, prior years adjusted to base year, and future years estimated without inflation), then year-dollars report (escalated or inflated dollars), and the discounted-dollar report (present value costs). If users selected the one subsystem for one-year option, HAZMAT generates costs for the 12 (predefined) cost elements.

Life-cycle stages covered

Raw materials acquisition
Manufacturing
Use/reuse/maintenance
Recycle/waste management

The acquisition phase includes information on acquiring various systems and subsystems. The operating and support phase uses information on the economic life of the subsystems, operating sites, etc. This corresponds to the use/reuse/maintenance stage of the life-cycle. The decommissioning phase uses information on the numbers and types of subsystems being decommissioned. This corresponds to the waste management stage of the life-cycle. Recycling is not taken into account.

Type of costs considered

Conventional
Potentially hidden
Contingent
External

The system has 12 different cost elements for actual costs of items such as facilities used, medical exams, etc. The cost elements are as follows:

· **Procurement:** actual purchase price of materials, plus transportation costs to site of use
· **Transportation:** costs of transporting materials from one location to another
· **Personal Protection:** three subelements include cost of personal protection equipment (including maintenance and support), cost of inefficiency (as a result of using that particular equipment), and the cost of dispensing the equipment
· **Management:** includes functions necessary to maintain oversight of materials (primarily a labor cost)
· **Training:** includes cost of training personnel
· **Handling:** cost of subdividing, labeling, and distributing materials, and the cost of controls placed on hazardous materials and their distribution
· **Potential Legal/Environmental Liability:** covers potential liability for violations (legal/environmental) of 12 Federal environmental acts or liability laws (related to amount of hazardous waste disposed). The cost has two subelements-one for contractor disposal and the other for air-home contaminants. The subelement for contractor disposal primarily covers third-party liability resulting from the transportation and storage at a waste disposal site. The subelement for air-home contaminants addresses provisions of the Clean Air Act.
· **Medical:** four subelements are occupational physical exams

(including lost time while physical is being administered), medical surveillance, cost associated with lost time due to illness/injury as a result of hazardous materials, and industrial hygiene surveys

· **Facilities:** cost of constructing and maintaining facilities, especially for hazardous materials and hazardous waste

· **Support Equipment:** cost of special equipment to handle hazardous materials and hazardous waste, including sensing devices and laboratory equipment

· **Emergency Response:** cost of emergency personnel and equipment

· **Disposal:** encompasses the cost to operate an Industrial wastewater Treatment Plant, where applicable, waste collection and handling, contractor disposal, and hazardous waste analysis and classification.

The third basic HAZMAT algorithm employs factors from the Pollution Prevention Benefits Manual (EPA230-R-89- 100), prepared by the EPA and ICF Incorporated, to estimate the Potential Legal/Environmental Liability cost element (and all subordinate cost elements). The developers of HAZMAT also cite a TASC Technical Information Memorandum as part of the basis for this algorithm.

Method of cost estimation

The methods used for calculating the costs are predicated on estimating the costs attributed to hazardous materials for the entire weapons system. Model algorithms are based on data collected for the estimates and are chemical specific. HAZMAT uses general and estimate information entered by users to do cost calculations. HAZMAT has algorithms for calculating costs under each of the categories explained above. The type of calculation depends on the choices entered by users. The algorithms may be the same or may vary depending on the phase under consideration. For instance, the algorithm for Transportation cost is the same for all subsystems and all phases. Algorithms to compute the cost of Training are different in different phases. Cost units available are thousands of dollars (default) or millions of dollars.

Generation of financial indicators

Net present value (NPV)
Payback period
Internal rate of return (IRR)
Benefits cost ratio
Others

HAZMAT calculates the Present Value (PV) of costs. The discount rate and the inflation table are entered in the first screen. The inflation table can be user defined, or HAZMAT defined. Users also enter the estimate base year and the cost input base year. Both these values have to be between the years 1975 and 2050, or else HAZMAT will default to 1993. Tables include raw and weighted indices.

Attributes

Output can be exported to a word processor or spreadsheet (after reports are generated). However, some reformatting may be necessary, and in a spreadsheet application, no arithmetic operations are retained. The system is menu-driven and is easy to understand and use.

All system-level parameter files are maintained in files or lists, rather than being hard coded. HAZMAT has permanent files and estimate files. The permanent files contain the entire HAZMAT database and cannot be changed by users. The estimate files contain the information on estimates (and data drawn from permanent files for the estimate). Any changes made to the data affects only the estimate files. HAZMAT cost algorithms are explained in the manual and are available for users to review.

Customer service is provided at HSC/EMP from 7.30 AM to 4.15 PM CST. HSC/EMP is also attempting to set up user groups in different geographical areas to enable trading of information on uses of the model, data from specific sources, and other information.

Limitations

The system is chemical specific and is not designed for estimating life-cycle costs of products. Users have to identify a system and subsystems that use one or more hazardous materials and make comparisons in that context. Users can enter up to three groups of chemicals with a maximum of 15 substances for the cost trade-off analysis. HAZMAT performance and run-time is highly dependent on the hardware used. The maximum number of years (schedules) that can be selected for any phase is 15 and not less than I year. The maximum cost is less than $10 trillion when cost units are expressed in millions of dollars. The system does not include external costs or list the environmental costs/benefits of using a particular hazardous substance. No additional cost categories may be added, and cost algorithms cannot be modified.

Basis for evaluation: The information provided here is derived from the HAZMAT user's manual and software provided by HCS/EMP in March, 1995.

Contact information: Available to any government facility/government contractor.
U.S. Air Force-HSC/EMP 8213 14th Street Brooks Air Force Base TX 78235-5246 Betty West 210-536-5121
TASC
2555 University Boulevard Fairborn, Ohio 45324

Hazardous Material Life-Cycle Cost Model: System Manager's Guide

Prepared by: Hermansen, L..A., J.A. Mooney, and W. Pugh.

Prepared for: Naval Health Research Center, Medical Information Systems and Operations Research Department, San Diego, CA

Application: Financial analysis: - Environmental impact analysis: - Waste management/P2: 1* Environmental cost listing/database: - Cost estimation: * Alternative product/process comparison: -

An instruction manual (29 pages) for a computer-based mathematical model that allows the computation of certain Lees associated with the use of hazardous materials in the construction, maintenance, and repair of U.S. Navy systems and facilities.

Development/publication date and updates: September 1992

Public availability: Available

Purpose and current use: This prototype model was developed to estimate the total life-cycle costs of using various hazardous materials in the construction, maintenance, and repair of U.S. Naval systems and facilities.

Cost information: Available at a cost of $17.50 from NTIS.

Contents and system requirements: Software, guidance manuals

Summary of methodology

The model estimates those costs derived from the need to protect the health and safety of workplace personnel and the need to protect the environment. The model considers three different exposure states: (1) no exposure, (2) personal exposure (the people involved are exposed over the personal exposure limit (PEL)), and (3) environmental exposure (unacceptable amounts of a hazardous material escape the confines of the facility).

The system allows users to use the materials, items, and costs provided by the system or enter project-specific materials, items, and costs. Cost sources include the Naval Health Research Center's Hazardous Materials Department, Environmental Planning and Compliance Department, Industrial Hygiene Department, Environmental Planning and Compliance Department, Office of Personnel Management and Department of Labor, as well as Medical Surveillance Procedures Manual, Industrial Safety Equipment Catalogue, and medical costs paid by San Diego County to private health care facilities.

Life-cycle stages covered

Raw materials acquisition
Manufacturing
Use/reuse/maintenance
Recycle/waste management

The model includes five life-cycle stages:

1. research and development
2. raw materials acquisition
3. processing/manufacturing (construction)
4. use (maintenance and repair)
5. waste management (final disposition)

Method of cost estimation

Users estimated costs for activities. The following methods of estimation are recommended in the 1992 fact sheet:

1. Differential accounting: only the future cash flows that would change if the pollution prevention project was implemented are accounted for.
2. Inclusion of labor hours.
3. Extended time frame: a 5-year period is recommended for analysis.
4. Modified financial criteria: a lower rate of return, lower discount rate, or shorter payback period is recommended because of intangible benefits such as reduced risk, improved quality, fewer sick days, or better public image.

Generation of financial indicators

Net present value (NPV)
Payback period
Internal rate of return (IRR)

Storage and transportation are also considered. Raw material acquisition has not been check-marked, because our definition of the term is different from the one implied here.

Type of costs considered

Conventional
Potentially hidden
Contingent
External

The methodology is used to compute only conventional and hidden costs (conventional costs of inputs and outputs) associated with hazardous material and waste management.

Method of cost estimation

The model requires users to specify values for a number of parameters including the type of material, phase of the life cycle, process, exposure type, and costs. The latter are given in a three-level hierarchical arrangement of cost item, cost element, and cost factor. One or more cost items are combined into a cost element, and one or more cost elements are combined into a cost factor. From the specification of these parameters total Lees are determined by a mathematical formula presented in the manual.

The manual provides no indication of actual costs of any of the contributing factors. However, appendix C indicates a number of sources from which these costs may be obtained. To compute the cost for activities taken at each exposure level, users first group them into cost factors. These factors include claims and compensation, disposal. engineering controls, fines and penalties, medical surveillance, medical treatment, permits and certification, personal protective equipment, procurement, spill containment and cleanup, storage, training, and workplace monitoring. Costing includes considering the probability the items will be used, the number of people involved, the number of days involved, and the quantity of the material.

Generation of financial indicators

Net present value (NPV)
Payback period
Internal rate of return (IRR)
Benefits cost ratio
Others

Attributes

This model considers the costs associated with research and development.

Limitations

The system is flexible, allowing the addition of processes, materials, etc., but users cannot add to the database. The model estimates only the Lees incurred in assuring the health and safety of workplace personnel and the protection of the environment. There is no information on specific materials, processes, levels of protection and costs in the report.

Basis for evaluation: The information provided here is based on a profile of the tool presented in Weitz et al. (1994). Subsequent reviews were done by incorporating information from a profile presented Hombach (1995).

Contact information: Available through NTIS, accession number: ADA269156. Hermansen, L.A., J.A. Mooney, and W. Pugh. *Hazardous Material Life-Cycle Cost Model Systems Managers Guide,* Naval Medical Research and Development Command, Naval Health Research Center, Report No. 92-17, September 1992, AD-A259156. For copies of the manual: National Technical Information Service (NTIS)
800-553-6847 703-487-4679 703-321-9038 703-321-8547 (FAX)

Life Cycle Costing Program-Version 2.0

Prepared by: Specialty Steel Industry of the United States (SSIUS)

Prepared for: Public use for comparing the life-cycle costs of stainless steel in comparison with other materials.

Application: financial analysis: · Environmental impact analysis: - Waste management/P2: - Environmental cost listing/database: - Cost estimation: - Alternative product/process comparison.

Cost benefit analysis of life-cycle costs for material selection. Life-cycle costs of materials other than stainless steel can also be compared, which can form a component of an environmental project. The system cannot take all life-cycle stages and costs into account. Not meant to compare project costs. Both large and small project data may be analyzed. However, cost details cannot be shown and the program cannot handle calculations that involve over 9 digits. There may be a practical limit on the time horizon that can be covered. Twenty to 100 years is typically used.

Development/publication date and updates: The program was developed in Switzerland (in EURO INOX-a sister organization of SSIUS) and modified by the International Chromium Development Association (South Africa) in - 1991. There are no plans for further development of the tool.

Public availability: Available

Purpose and current use: The software was developed as part of the SSIUS's effort to promote the use of stainless steel. Many users consider stainless steel to be an expensive option, but the economics can be different if other life-cycle stages are included in the analysis (other than capital costs). However, the vendor claims that the tool is suitable for designers and material specifiers dealing with a wide range of applications.

Cost information: The program is available free of charge for anyone in a position to use it for materials selection. There is (-) a

$10 charge for people who plan to use the program as a sales tool.

Contents and system requirements: The minimum system requirements are an IBM (or compatible) PC, DOS version 2.2 and 640 K RAM.

Summary of methodology

This LCC program assesses the potential long-term costs and benefits against short-term expediency, when making materials selection. The program compares likely lifetime costs of up to three material alternatives in present value terms. The program functions include the following:

- sensitivity analysis (what-if analysis) on the primary input variables
- on-screen display of cost summaries
- printouts of cost summaries, detailed analysis, and sensitivity analysis

The sensitivity analysis can view effects on total costs of changing the base case values of primary input variables by selected percentages. The results are displayed in Mu's (monetary units). The tool does not allow any qualitative information to be entered. The program converts all costs (other than initial costs) to present values, to calculate total life-cycle costs in present value terms. The software does not limit the length of the life cycle as long as the total LCC figure does not exceed 9 digits.

Life-cycle stages covered

Raw material acquisition
Manufacturing stage
Use/reuse/maintenance
Recycle/waste management

The system is not designed to consider all than one life-cycle stages. A different definition of life-cycle costing is used here. It is defined as the total costs of a project from the initial outlay through all related future costs (maintenance, replacement and other operations) that may be incurred as a consequence of the initial investment decision. Thus it focuses on just one life cycle stage-typically the use/reuse of a particular material and the initial cost.

Type of costs considered

Conventional
Potentially hidden
Contingent
External

The tool is designed to consider conventional and hidden costs (e.g., maintenance, replacement) associated with an investment decision. The system does not consider all potentially hidden, contingent, or external costs associated with different material alternatives.

Since users enter only total values, they can make operating cost inputs more comprehensive (if they so desire) by including waste management and other environmental costs. Alternatively, they may be able to estimate the percentage impacts, for instance, of liability or waste management costs on the operating costs. These impacts can be analyzed in the sensitivity analysis (limited, however, to values between 1% and 99%). The program allows costs to be broken down only to a certain extent, which may force users to lump certain cost categories together. Users have to set up a single value for downtime periods for maintenance and repair.

Method of cost estimation

The program calculates life-cycle costs as :

Life-Cycle Cost (LCC) = Initial materials acquisition costs + Initial material installation and fabrication costs + Present Value (PV) of operating and maintenance costs + PV of lost production costs during downtime + PV of replacement materials costs.

Users first need to enter the following information:

1. collected data on material options (up to three) for a specific application
2. values for rates and duration (any)
3. initial costs
4. operating costs

Details need to be entered for downtime per maintenance/replacement event, value of lost production per day and initial total (per unit) material costs, fabrication and installation costs, other installation costs, maintenance costs, replacement costs, and annual material-related costs (indirect operating costs). Users also need to estimate the number of maintenance/replacement events and elapsed time between them. They can take into account the residual or scrap value of material after replacement.

The program calculates the following:

1. Real interest rate (corrected for inflation), using the formula:
$r = (n-q)/(1+q)$ where
n = cost of capital (nominal bank rate) q = inflation rate This is the actual discount rate that is used for the PV calculations.
2. Cost of lost production-in present value terms

It performs sensitivity analysis using percentages that can be selected between 1% and 99%.

Costs should be entered as per-unit values, which are later multiplied by the number of units to obtain overall costs. The inflation rate is always assumed to be less than the cost of capital.

Generation of financial indicators

Net present value (NPV)
Payback period
Internal rate of return (IRR)
Benefits cost ratio
Others

The program discounts all cash flows to get PVs, which can then be used as a basis for comparison. The program takes material, fabrication, and installation and other installation costs as initial costs and therefore does not discount them. These are taken as total initial costs. Using the real interest rate and the PV formula, the program calculates PVs for maintenance costs, replacement costs, cost of lost production, and annual material related costs. However, users cannot enter values for benefits. Thus the program does not compute NPVs.

Attributes

Users do not have to enter data for all input prompts, although the study would be more comprehensive if they did. There really is no othe avenue for modifying or customizing the program. The guide warns that the calculated real interest rate can sometimes be too high for projects with long time periods. Appropriate adjustments need to be made to the chosen cost of capital. The program has a help menu that assists with input definitions and program functions. The program gives an error message for inconsistent values.

Limitations

The program is suitable for preliminary analysis of material alternatives. The program does not support cost estimation or the input of detailed cost data. Total costs have to be estimated elsewhere before input into the program, which performs the financial analysis. Also, the program only calculates the PV of costs not net benefits. Users receive no information on estimating the life-cycle costs included (e.g., maintenance or replacement costs). However, the tool can be utilized along with cost estimating systems to perform financial analyses.

The program does not export/import data from elsewhere. This is not a major limitation because it does not call for large inputs. The program inputs for material costs are divided into 1) plate, sheet; 2) pipe, tube, fittings; and 3) bar and other. Thus the input prompts are tailored toward selecting steel and related substitutes. This may limit the types of materials that users want to consider, unless they choose to ignore the category names. The program cannot support calculations that have greater than 9 digits. This can be a problem with, for example, long life cycles and high operating costs. This is not meant to be a commercial product and hence no user support is provided (beyond minor inquiries).

Basis for evaluation: Review of disk and information. Additional information obtained through telephone conversation with Brian Leslie in March 1995.

Contact information: Available, but not a commercial product
Brian Leslie, Director of Market Development Specialty Steel Industry of the United States 3050 K Street, N.W. Washington, DC 20007
202-342-8630 202-338-5534 (FAX)

Models Developed for the Total System Life-Cycle Cost Analysis

Prepared by: Roy F. Weston Inc.

Prepared for: U.S. Department of Energy, Office of Civilian Radioactive Waste Management (OCRWM)

Application: Financial analysis: - Environmental impact analysis: - Waste management/P2: / Environmental cost listing/database: - Cost estimation: Alternative product/process comparison: -

Description of the models used in the 1989 Total System Life-Cycle Cost (TSLCC) analysis, which is an annual study of costs for the radioactive waste disposal program administered by the OCRWM. The TSLCC analysis establishes the long-term financial plan for the OCRWM. This integrated set of models was developed by OCRWM to provide an analytical basis for establishing this financial plan. The TSLCC models have also been used to prepare cost estimates contained in several other DOE documents, and have applications in cases involving variations in many types of waste management system parameters. Any applications would have to be reviewed on a case-by-case basis.

Development/publication date and updates: 1990. Published in 'Transactions of the American Nuclear Society', Volume 60,1989. The report, "Analysis of Total System Life Cycle Cost for the Civilian Radioactive Waste Management Program" is an annual publication.

Public availability: This would need to be established from the DOE.

Purpose and current use: The total system life-cycle cost analysis for the DOE Civilian Radioactive Waste Management Program is an ongoing activity that determines whether the revenue- producing mechanism established by the Nuclear Waste Policy Act of 1982-a fee levied on the producers of nuclear power-is sufficient to cover the cost of the program. The use of these models formed the basis for the annual report to Congress. These models are not currently being used by the DOE.

Cost information: Cost of the models cannot be determined at this point. The TSLCC report is available free of charge through OSTI.

Contents and system requirements: Software packages. The models used for total-system cost analysis consist of a series of computer programs operating in both mainframe and microcomputer environments, which are integrated through a file management system.

Summary of methodology

Separate computer programs have been designed for the major cost components of the waste management system: each program is structured to accommodate the best available data and the most recent design information relevant to performing the cost analysis. The following sections describe how the models integrate and function sequentially.

· Spent nuclear fuel logistics **(WASTES)** model and the defense waste logistics **(DHLWLOG)** model analysis: together determine the waste transfer rates between system components and the characteristics of the waste relative to the cost algorithms (e.g., waste age, burnup). The WASTES model also calculates from-reactor spent-fuel transportation costs.
· Waste Package Logistics **(WPLOG)** model: processes the output

from the WASTES model to determine the number of waste packages to be placed in the repository.
· **URCOST** for the subsurface costs and **REPCOST** for the surface facility costs: two large spreadsheet models that use the information from WPLOG to estimate the repository construction, operating, and decommissioning costs.
· **MRSCOST:** a spreadsheet that uses the WPLOG output is as input for estimating the monitored retrievable storage facility costs.
· **ARSCOST:** a cost routine that uses the output of the WASTES model for storage requirements to estimate at-reactor storage costs.
· **DECOST:** name for a series of spreadsheets that estimate development and evaluation costs.
· **WITCOM** model: takes in output from all of the above programs and performs several cost functions to yield the total-system cost.

Life-cycle stages covered

Raw materials acquisition
Manufacturing
Use/reuse/maintenance
Recycle/waste management

The following life-cycle information is considered.

· transportation/packaging
· waste management (development and evaluation costs, repository costs, costs for a monitored retrievable facility, costs of at reactor storage)

Type of costs considered

Conventional
Potentially hidden
Contingent
External

The method are used to estimate conventional costs of inputs and outputs and considers some aspects of hidden costs.

Method of cost estimation

Cost data is from various DOE sources and reports.

Generation of financial indicators

Net present value (NPV)
Payback period
Internal rate of return (IRR)
Benefits cost ratio
Others

Attributes

The modular programming system facilitates updating specific units when appropriate. Some of the major types of parameters that have been analyzed using these models include alternative engineering designs for facilities, alternative facility locations, alternative waste forms, alternative waste acceptance schedules, and others.

Limitations

The systems are not applied for total product life-cycle cost analysis, as all the life-cycle stages are not covered. External and contingent costs are not included in the analysis (although they potentially could be).

Basis for evaluation: The information provided here is based on a profile of the tool presented in Weitz at at. (1994). Subsequent reviews were done by incorporating information obtained through telephone communications with Roy Weston Inc. in May, 1995

Contact information: Roy F. Weston, Inc. 955 L'ensant Plaza, SW, 8th floor Washington. DC 20024
202-646-6654
For copies of the report "Analysis of the Total System Life Cycle Cost for the Civilian Radioactive Waste Management Program," DOE/RW/0236, May 1989:
Office of Scientific and Technical Information (OSTI) 615-576-8401
Copies of the 1985,1986,1987, and 1990 reports are also available through OSTI.
Office of Civilian Radioactive Waste Management (OCRWM) 800-225-6972

TEAM™ (Tools for Environmental Analysis and Management)

Prepared by: Ecobalance, Inc.

Prepared for: Internal and commercial use

Application: Financial analysis: - Environmental impact analysis: - Waste management/P2: fr* Environmental cost listing/database: - Cost estimation: - Alternative product/process comparison.
Life Cycle Cost and Life Cycle Assessment software

Development/publication date and updates: TEAM™ has been developed for Ecobalance by Net-ID, Inc., a California-based company specializing in advanced computer science and computations in hierarchical networks. Development started in 1992. Version 1.15 released in May, 1995.

Public availability: Available

Purpose and current use: LCC and LCA. The model has been licensed to approximately 40 companies and government agencies worldwide.

Cost information: Contact Ecobalance

Contents and system requirements: Windows-based program, developed in C++ with object-programming and an open architecture. Windows NT version available end of 1995. For the current version, the requirements are: IBM-compatible PC(486 or higher). Windows 3.1 or 3.11, 8 MB RAM, 40 MB hard drive (recommended use, program needs 3 MB).

Summary of methodology

TEAM™ is a Life Cycle Cost (LCC) and a Life Cycle Assessment (LCA) calculation tool. It is also a comprehensive database on processes and materials, DEAM™ (Data for Environmental Analysis and Management). The information provided in the database is usually referring to the 1988-1994 period. In all cases, the origin, geographical and technical representativeness of the data, and the underlying assumptions are mentioned.

Users can build their own database to use in conjunction with the DEAM™ database and model any type of industrial system. The software generates life cycle inventories and life-cycle cost results, including sensitivity analysis. Users deal with operations or "modules" (industrial processes, transport, commercial activities, etc.), and systems (set of operations linked together). Modules and systems have inputs and outputs, corresponding to costs, natural resources, intermediate or finished products, energy sources, wastes and environmental releases. Modules are connected through their inputs and outputs (a module accepts any number of connections). In addition to storing numerical inventory values, one can store information on the source of information, uncertainty, variability, measurement techniques, etc. Modules and systems created for a given project can be stored to be reused for other purposes.

A wide variety of projects can be performed in TEAM™, ranging from simple life cycle estimates (built in a few days) to complex projects. The largest system built to date gathers approximately 2,200 linked operations in several layers of nested sub-systems.

Life-cycle stages covered

Raw material acquisition
Manufacturing stage
Use/reuse/maintenance
Recycle/waste management

All life cycle stages, including transportation, are covered.

Type of costs considered

Conventional
Potentially hidden
Contingent
External

The software does not come with a default format that address any of these cost categories. However, the software has been designed with the intent to provide users with the ability to incorporate cost information into life cycle analysis. An indefinite number of user-defined cost categories can be added to the model. In particular, conventional, potentially hidden, and contingent costs can be addressed. Consideration of external costs is not recommended due to the uncertainty involved in doing so.

Generation of financial indicators

Net present value (NPV)
Payback period
Internal rate of return (IRR)
Benefits cost ratio
Others

User-defined formulas allows one to create specific indicators. All classic mathematical functions are available.

Attributes

TEAM™ permits very complex systems to be analyzed while keeping the interaction with users simple. Features include: full documentation of data (origin, uncertainty, etc.) and assumptions in dedicated fields, ability to protect confidential data, and use of variables and formulas in modeling systems, allowing one to carry out what-if analyses. Key parameters (e.g., design parameters, recycling rates, transportation distances) can be centralized in a single monitoring panel. The embedded database contains information on a variety of commodity materials and processes, including pulp and paper, petrochemicals, plastics, inorganic chemicals, steel, aluminum, glass, energy conversion, transportation and waste management.

Annual and indefinite maintenance and database updation contracts can be provided, draining classes and user- support is available. The software comes with on-line help. The program supports the full range of Windows capabilities.

Limitations

TEAM™ does not have any default format. It essentially provides users with the modeling tools using which they can build their analysis. Step by step instructions are not provided by the software. The software is flexible enough to include any type of costs, but there is no specific guidance provided to users on what types of costs to consider.

Basis for evaluation: Sample profile prepared by Ecobalance Inc..

Contact information: Available

Ecobalance, Inc. I Church Street, Suite 700 Rockville, MD 20850 Remi Coulon (Technical Manager) 301-309-0800 301-309-1579 (FAX) E-mail: 100135.2667 @ Compuserve.com

Tellus Packaging Study

Prepared by: Tellus Institute

Prepared for: The Council of State Governments, U.S. Environmental Protection Agency New Jersey Department of Environmental Protection and Energy

Application: financial analysis: - Environmental impact analysis: * Waste management/P2: * Environmental cost listing/database: - Cost estimation: - Alternative product/process comparison.

Assessment of the impacts of production and disposal of packaging and public policy measures to alter its mix.

Development/publication date and updates: May 1992. No updates

Public availability: Available

Purpose and current use: Analysis of all major packaging materials. Tellus uses the system for their LCA and other environmental policy work. Tellus has sold a number of copies of the study to industry, universities, non-profits, and government organizations be used for various purposes, including support for policy decisions and frameworks.

Cost information: The complete study is available at a price of $295 for non-profit organizations, and $495 for for-profit organizations. The executive summary is available for $35 and $55 for non-profit and for-profit organizations, respectively.

Contents and system requirements: Detailed analysis of all major packaging materials, including an evaluation of the relative importance of different environmental impacts and pollution control costs. Case studies of packages.

Summary of methodology

The Tellus approach uses the marginal cost of pollution control for regulated and non-regulated pollutants to evaluate the environmental cost of production associated with twenty packaging materials, including several comparisons of virgin and recycled streams.

Tellus works within the SETAC/EPA LCA framework-consisting of a life-cycle inventory, impact assessment, and improvement assessment. The Tellus inventory report on packaging material production processes coincides with the inventory component of life-cycle analysis. Ranking health effects and pricing them are part of the life-cycle impact assessment.

Life-cycle stages covered

Raw materials acquisition
Manufacturing
Use/reuse/maintenance
Recycle/waste management

The Tellus study evaluates environmental impacts/costs for a portion of the life cycle and includes information on

· raw materials extraction/processing
· packaging material production and
· disposal of packaging material.

The Tellus study omitted the following life-cycle stages:

· package forming
· filling
· transportation
· shipping materials (e.g., pallets, wrappings)

Type of costs considered

Conventional
Potentially hidden
Contingent
External

The costs considered are environmental costs of production, conventional (monetary) costs of disposal, and environmental costs of disposal.

Environmental costs of production

· Uses the marginal cost of pollution control (MCPC) as a proxy for estimating cost

· MCPC values pollutants at the price society is willing to pay for pollution controls

· Pollutant "prices" for regulated pollutants established using their MCPCs, and pollutant prices for non- regulated pollutants established using the MCPC of lead as a benchmark

· Total costs for each packaging material determined by multiplying each pollutant's price by the amount of each pollutant released to the environment per ton of material produced.

Conventional costs of disposal

· Measurement of the marginal cost of handling various packaging materials.

· Includes costs of landfilling, incineration, and recycling · Uses New Jersey's solid waste system as baseline model

Environmental costs of disposal

· Costs were determined using the pollutant price, as in estimating production costs, multiplied by the amount of pollutant released per disposal method:

-For landfills, the leachate pollutants per ton of landfilled waste were estimated and these were added to the waste stream components based on the quantities and chemical composition of wastes.

-For incinerators, the air pollutants and ash produced by burning a ton of waste were estimated and these were added to the waste stream components based on the quantities and chemical composition of wastes.

-For recycling, the air pollutants associated with recycling a ton of recycled waste were estimated and these were added to the waste

stream components based on the quantities of waste. One published study was used that measured ambient air pollution levels at a fairly small recycling facility. The reason why only one data set was used was the paucity of published data on recycling facility emissions.
-For landfilling, incineration and recycling, the air pollutants associated with collecting a ton of waste were estimated, and these were added to the waste stream components based on the quantities and volumes of wastes.

Method of cost estimation

Full cost of packaging production and disposal =
Environmental costs (pollutant prices) of packaging production +
Conventional (monetary) costs of the solid waste system +
Environmental costs (pollutant prices) of the solid waste system

Generation of financial indicators

Net present value (NPV)
Payback period
Internal rate of return (IRR)
Benefits cost ratio
Others

Attributes

Most applicable aspects of the Tellus study are the approaches used to place monetary values on environmental impacts. Although city production and disposal stages are considered by Tellus, the use of social WTP (as given by the MCPC) as a proxy for pollutant prices may be built upon, refined, and/or extended to all life-cycle stages.

Limitations

The conventional costs of packaging production are not taken into account. All the life-cycle stages are not included in the analysis.

Basis for evaluation: The information provided here is based on a profile of the tool presented in Weitz et al. (1994). Subsequent reviews were done by incorporating updated product information through telephone communication with Tellus in May 1995.

Contact information: Tellus Institute 11 Arlington Street Boston, MA 02116
Karen Shapiro Brian Zuckerman
617-266-5400 617-266-8303 (FAX)

ABBREVIATED PROFILES : LIFE-CYCLE COST ANALYSIS

Abridged LCA Matrix

AT&T developed an abridged matrix approach for LCA to facilitate semi-quantitative, efficient assessments of complex manufactured products or industrial manufacturing processes. This is in recognition of the complexities of a complete LCA coupled with the difficulties in relating inventory analysis to impact and improvement analysis. The abridged approach attempts to allow firms to do rapid LCAs, for example, 2 days for a typical product or one week for a typical facility.

The matrix uses a scoring system for comparing options while focusing on those design and implementation aspects whose modification could most dramatically affect the rating. Five life-cycle stages are taken into account:

Stage I Premanufacturing Stage 2 Manufacturing Stage 3 Product delivery (including installation) Stage 4 Customer stage Stages Termination of product's life. Each stage is controlled by different parties, for instance, the customer stage is not directly controlled by the manufacturer, but is influenced by how products are designed. A 5 x 5 assessment matrix, called the Environmentally Responsible Product Assessment Matrix, is recommended. One dimension is the life-cycle stage, and the other is the environmental concern. The ratings are from 0 (lowest impact) to 4 (highest impact), with higher numbers reflecting more positive evaluations. The assessment is subjective and the assessors need to have adequate information to allow comparative ratings and evaluations on a variety of characteristics. For example, products can be compared against materials choice, energy use, solid residue, liquid residue, and gas residue, in each of the life-cycle stages. According to some AT&T evaluations, these informal assessments vary only slightly among different groups of assessors. The matrix approach can be made more detailed by applying weighting factors to the matrix elements.

This approach does not take cost considerations into account, nor does it consider all the environmental impacts. The subjective nature of the evaluations will cause some value judgments and the accuracy of the assessment is dependent on the expertise and knowledge of the assessor's). However, the approach does provide organizations with an approach to take to account environmental considerations that would otherwise be ignored.

Contact Information: AT&T Bell Laboratories Room ID 349
Murray Hill, NJ 07974
Tom Graedel 908-582-5420

The description of the model can also be found in a textbook called 'Industrial Ecology' by Tom Graedel and Brad Allenby. This book is available for $37 from Prentice Hall Publishing (515-284-6761), ISPNO-13-125238-0.

Basis for evaluation: Graedel T.E., Allenby B.R., Comrie P.R. "Matrix Approaches to Abridged Life Cycle Assessment". Environmental Science and Technology. Vol.29, No.3, 1995.

ASEAM (A Simplified Energy Analysis Method)

The program is targeted to Federal Energy managers, contractors on Federal building construction projects. State and local government energy managers, and other contractors. ASEAM is a method for calculating energy consumption of residential and simple commercial buildings.

ASEAM uses the ASHRAE (American Society of Heating, Refrigerating, and Air-conditioning Engineers) energy file closest to the area to calculate annual and monthly energy requirements. It permits evaluating options for energy conservation. ASEAM calculates monthly and annual energy consumption for buildings and allows calculations of energy savings through options such as more insulation, higher energy heat pumps, and double glass windows. By feeding in costs of these options and using BLCC to determine life-cycle costs, the economics of energy conservation can be evaluated. ASHRAE-maintained national weather data are included with the software.

It is expected that future upgrades will include manufacture-specific systems replacement and compatibility with hourly analysis techniques. The program can be obtained free of charge. ASEAM is updated every few years. The new release of ASEAM (version 5.0) is expected later in the year.

Minimum Hardware: IBM PC with a floppy disk or equivalent; will not run on a Macintosh. The system cannot be accessed by multiple users on a network.

USDOE, Federal Energy Management Program 1000 Independence Ave., SW Washington, DC 20585
Dean DeVine EE-92 202-586-6784

Purchase of software: Advanced Sciences, Inc. 1525 Wilson Blvd, Suite 1200 Arlington, VA 22209 703-243-4900

Basis for evaluation: Based on Watson (1995).

BLCC (Building Life Cycle Cost Program)

The program was developed to support LCC evaluations, which are required for all Federal building construction projects. The program is also available to the private sector. BLCC is used to evaluate life-cycle costs for a building to evaluate the impact of energy and water cost saving measures on life-cycle costs. It can provide economic analysis for different project evaluation environments, including federal government and private sector projects with taxes, and general projects that do not require tax analysis or specific LCC guidelines. BLCC allows users to compare life-cycle costs for various energy and water conservation measures with user-supplied energy and/or water usage rates and prices. It allows users to decide between construction options (e.g., whether to use a higher initial cost energy system to reduce energy costs throughout the building life). Costs of different options can be compared. BLCC computes life-cycle costs, savings-to-investment ratio, net savings, internal rate of return, and cash flow analysis for project alternatives. Up to 99 alternative designs can be evaluated simultaneously. Six reports including reports by project alternative and reports comparing project alternatives can be generated. BLCC complies with American Society for Testing and Materials (ASTM) standards related to building economics. Other software that is included on the BLCC diskette:

- **Quick Input** (01) **program:** Convenient for solving relatively

simple LCC problems and preliminary economic evaluations of multiple design alternatives.. It can generate BLCC input data files.
· **DISCOUNT:** A stand-alone program that performs the individual discounting operations required in LCCA, and generates various financial indicators. This program can access the DOE projections of energy price increases used in the BLCC program. It is useful for evaluating individual components of a LCC problem that do not require comprehensive analysis.

ERATES is another stand-alone program that can be used for generating block-rate, time-of-use rate and demand- rate schedules for energy prices, which are further used to compute monthly and annual electricity costs. BLCC and 01 can also compute annual and kWh and kW demand charges with monthly block-rate schedules imported from ERATES. This program is included on a separate diskette.

BLCC and 012.0 are compatible with ASEAM (for whole building analysis) as well as with Motor Master (for motor replacements). The Federal version comes with the required federal discount rates and is revised annually. The private sector version allows user-defined discount rates. BLCC is menu-driven (not Windows-based) with data entry screens. There is no cost database, and users need to input the required cost information, including capital and O&M costs, and utility costs. Users can define up to six components in a project, which may enable the incorporation of additional cost categories, such as less tangible costs.

All these programs were developed under sponsorship of the Federal Energy Management Program. They all include user's guides. The BLCC guide includes information on 01. The program is free to Federal users, and private users pay some nominal charges. BLCC is updated every October. The current version is 4.23-95. BLCC Version 4.3-96 will include enhancements such as the ability to work with DOC'S ECIP (Energy Conservation Investment Program) reporting requirements, and an improved ability for dealing with air pollution emissions (related to energy use).

Minimum Hardware: IBM PC with floppy disk or equivalent. The software is on two 31/2 inch, 1.4 MB disks. No special hardware or graphics capabilities are needed.

Software purchase by Federal facilities and contractors

Advanced Sciences, Inc. (Arlington, VA): 703-243-4900

Software purchase by other groups

Energy Information Services (St. Johnsbury, VT): 802-748-5148 FlowSoft (St. Charles. MO): 314-441-1022 NTIS Accession #: PB91-507970/XAB

Information on DOE sponsored LCC workshops conducted by NIST: US Department of Commerce, National Institute of Standards and Technologies Computing and Applied Mathematics Laboratory, Office of Applied Economics Gaithersburg, Maryland 20899 301-975-6132 Steve Petersen 301-975-6136

Instructional videos produced by HIST in the series "Least-Cost Energy Decisions for Buildings": Video Transfer Inc. (Rockville, MD): 301-881-0270

Basis for evaluation: Based on Watson (1995), and on information received from NIST in July, 1995.

FEDS Release 3.0

FEDS is a tool for assessing and analyzing energy efficiency in multiple buildings and at a single site. It is targeted an analysis tool for a wide range of energy systems professionals including energy managers, facility managers, architect-engineers, utility planners, and energy consultants. FEDS uses data entered by users (data on energy, building information, location, consumption, energy price data, building types, characteristics, operating hours and utility incentives) to develop a building prototype and inferred engineering parameters. These inferred parameters can be changed by users if better information is available.

FEDS helps in identifying energy improvements that offer maximum savings. It identifies and selects minimum life- cycle cost retrofits, determines payback periods, and enables users to prioritize options. FEDS includes a current database of hundreds of proven retrofit technologies, such as retrofits for lighting, heating, cooling, etc. Other features include an interactive format, life-cycle cost optimization, technology and fuel independence (no technology type or fuel is favored over the other), peak-demand modeling, and analysis using site specific data or projections. FEDS includes the same energy cost escalation rates as BLCC (Building Life-Cycle Cost program), and meets the life-cycle cost requirements of 10 CFR Part 436 used in BLCC. FEDS 3.0 includes enhancements such as new retrofit options, the latest discount rate, and an expanded and updated database.

FEDS is available for use on federally funded projects, or for those projects funded or directly performed by an agency of a State Government. The software is not currently available for private users, although efforts are under way to license the tool for commercial use.

Minimum Hardware FEDS 3.0 runs on Windows 3.1, or greater. U.S. Department of Energy, Federal Energy Management Program Washington, DC 20585 K. Dean DeVine, P.E. (202)586-6784 dean.devine@hq.doe.gov
Rosemarie Bartlett (FEDS Software Administrator) Pacific Northwest Laboratory PO Box 999 Richland, WA 99352 509-375-6606 509-375-3970 (FAX) r_bartlett@pnl.gov

Basis for evaluation: Based on information received from PNL.

"This page purposely left blank"

CHAPTER 10. WASTE REDUCTION SYSTEMS

Avoidable Waste Management Costs at DOE Facilities

Prepared by: Sandia National Laboratories (SNL), Waste Minimization/Pollution Prevention Program

Prepared for: U.S. DOE under contract DE-ACO4-76DP00789

Application: Financial analysis: - Environmental impact analysis: - Waste management/P2: Environmental cost listing/database: - Cost estimation: - Alternative product/process comparison:

This method is designed specifically for use at SNL/NM, and is tailored from process waste assessment procedures recommended by DOE. With appropriate modifications, the methodologies discussed can have applications in other facilities and firms as well.

Development/publication date and updates: November 1992 (updated January 1993). Completed in early 1994. Sandia Report: SAND92-2343, Version 1.0

Public availability: Not available

Purpose and current use: The manual was prepared by the waste minimization/pollution prevention program, as part of SNL's commitment to DOE to significantly reduce the generation of hazardous wastes. This method identifies and evaluates waste minimization projects at SNL. The manual was used to a limited extent at SNL.

Cost information: Not applicable. The cost of the conference paper (through NTIS) is $17.50 plus $4 for shipping and handling.

Contents and system requirements: Worksheets, data forms, software for chemical information/MSDS, and Environmental Health and Safety reporting. A software product was developed for phase 4 of the method: selection of optimal projects.

Summary of methodology

The method consists of four phases:

1. Process definition
2. Process characterization
3. Waste minimization opportunities assessment
4. Project selection, optimization, implementation, and tracking

Process definition provides qualitative information concerning the nature of processes and operations that contribute to generating wastes. This phase calls for describing overall operations, describing specific processes or operations that consume or generate wastes, developing a schematic flow diagram for these processes, and documenting previous waste minimization activities associated with these processes. Process characterization collects information concerning the quantity and nature of the wastes and other pollutants generated. Waste minimization opportunities assessment identifies and evaluates various alternatives for reducing or eliminating wastes generated in a process. Process descriptions (e.g., flow diagrams, system interactions, and mass balances) that are more detailed than those produced in the previous phases may be necessary to identify waste minimization opportunities. The final phase- project selection, optimization, implementation and tracking-involves managing and evaluating the chosen options.

Phases I and 2 are fully documented with blank and completed-example worksheets. Only overall summaries were available for Phases 3 and 4 as of the January 1993 version. Since an updated copy of the report is not available, there is no detailed information on these phases.

Life-cycle stages covered

Raw materials acquisition
Manufacturing
Use/reuse/maintenance
Recycle/waste management

Type of costs considered

Conventional
Potentially hidden
Contingent
External

Cost and technical analyses were done as part of the waste minimization opportunities assessment phase. These were based on EPA guidance. There is no information on the cost categories that were included in the analysis.

Method of cost estimation

There is no information available on the methods employed for cost estimation.

Generation of financial indicators

Net present value (NPV)
Payback period
Internal rate of return (IRR)
Benefits cost ratio
Others

Information not available.

Attributes

A paper presented at the Waste Minimization/Pollution Prevention conference, called "Waste Minimization/Pollution Prevention at R&D Facilities: Implementing the SNL/NM Process Waste Assessment Program" is available through NTIS.

Limitations

This document provides background for evaluating SNL processes. It will not have direct application in other areas.

Basis for evaluation: The information provided here is based on a profile of the tool presented in Weitz et al. (1994). Subsequent reviews were done by incorporating information obtained through telephone communications with SNL in May, 1995.

Contact information: The report is not available for public distribution. A related conference paper called "Waste Minimization/Pollution Prevention at R&D Facilities: Implementing the SNL/NM Process Waste Assessment Program," is available through the National Technical Information Service (NTIS): report number SAND-93-0466C, NTIS accession number: DE 93012860. Sandia National Laboratories Org. 6641, M/S 0718 Albuquerque, NM 87195-5800
Ed Kjeldgaard 505-845-8011 505-844-0116 (FAX)
National Technical Information Service (NTIS) 703-487-4660 800-553-6847

SRS Waste Cost Analysis

Prepared by: Advanced Planning and Analysis, Westinghouse Savannah River Company

Prepared for: Westinghouse Savannah River Company, Savannah River Site (SRS)

Application: financial analysis: - Environmental impact analysis: - Waste management/P2: >* Environmental cost listing/database: - Cost estimation: * Alternative product/process comparison: -

This method provides waste cost data and modeling techniques used to derive life-cycle costs for input for operational studies and evaluation of alternatives. The methodology uses a combination of cost/budget and bottoms-up task-based cost estimation.

Development/publication date and updates: May 15,1992. An updated document is called FY-1 993 SRS Waste Cost Analysis: report number WSRC-RP-93-942, Revision I.

Public availability: Available

Purpose and current use: The purpose of this report is to provide a life-cycle cost (that is, from the point of waste generation to disposal) for a unit volume of waste generated in SRS in the year 1992. The methodology is not currently being used at SRS. They are currently considering the use of a new, activity or task-based methodology.

Cost information: The document is not commercially available.

Contents and system requirements: Method for cost calculations, cost figures for SRS. The report is also available on disk (Macintosh) and includes two Microsoft Excel spreadsheets used in documenting and calculating life-cycle costs.

Summary of methodology

Issues that can be resolved using the data and techniques suggested by the method include waste minimization alternatives, cost charge-back to waste generators, legacy costs for long-term waste management, and comparison of options in decontamination and decommissioning.

Life-cycle stages covered

Raw materials acquisition
Manufacturing
Use/reuse/maintenance
Recycle/waste management

Costs are calculated for only part of the life-cycle: from waste generation to final disposal, including surveillance of the waste after isolation from the environment and closure costs for facilities, burial grounds, etc. The manufacturing stage is considered to the extent that costs of waste minimization activities are taken into account.

Type of costs considered

Conventional
Potentially hidden
Contingent
External

Only conventional and hidden costs of inputs and outputs are considered. These include costs of handling, packaging,

characterization, storage, treatment, disposal, pre and post-closure, environmental health, safety, quality assurance, site services, storage, and disposal.
Inputs and outputs include the following:

· generator costs
· facility operation
· transportation
· capital equipment
· facility upgrades
· monitoring
· support services
· technical center
· replacement facilities

Specific items not included in the waste costs include the following:

· raw material costs of waste
· existing facility capital costs
· DOE costs
· decontamination and decommissioning/corrective action costs
· infrastructure capital costs
· liability/risk costs
· recycling, reclamation activities

Method of cost estimation

Costs are divided into fixed and variable components. Variable costs vary in direct proportion to changes in the level of activity or the quantity of waste for treatment or disposal. Fixed costs remain constant regardless of changes in throughput within a reasonable range of operation of the waste facilities. Sunk costs are omitted. Current and future waste costs, based on the fiscal year 1992 waste generation and handling rates, have been compiled for all types of waste generated at SRS. For example, the cost categories in the case of sanitary waste (relatively similar for the other cases as well) are generator costs, transportation costs, waste management costs, closure costs, and monitoring costs. Cost estimations are done on the basis of a number of assumptions that are explained in the manual. The model used for life-cycle cost analysis is as follows.

Total life-cycle cost for a waste category = Fixed cost + * (Unit variable cost for each waste stream within a waste category) * (Forecasted volume for each waste stream)

The first in - first out (FIFO) accounting principal was used to simplify cost calculations when programs associated with managing a waste stream were predicted to be relatively constant. The current and future scheduled operating budgets were assumed to represent the level of funding required to effectively support the activities associated with managing waste.

Generation of financial indicators

Net present value (NPV)
Payback period
Internal rate of return (IRR)
Benefits cost ratio
Others

Attributes

Values for the SRS site provide a basis for comparing sanitary waste, solid low-level waste, liquid low-level waste, transuranic waste, hazardous waste, mixed waste, and high-level waste. The manual itself does not guide users toward decision-making or generating alternatives. However, by estimating waste management costs, it gives users important data that can be used for making such decisions. A number of different assumptions are used to calculate fixed and variable costs. These assumptions were discussed in the context of waste type. The information on assumptions can be helpful to other users.

Limitations

The manual needs to be revised periodically to reflect the changes in the cost database. The authors had envisioned annual updates of the document. However, SRS is currently considering moving to a new methodology. The report has been produced for SRS and is based on their forecasts, waste facilities, and waste characteristics. Outside users cannot use the manual "as is" to support their waste management plan. It can only be used as a guide for cost estimation or the preparation of a similar guide specific to another company or area, with the relevant assumptions and objectives.

The manual is restricted to conventional and some hidden cost calculations and does not attempt to estimate either liability costs or external costs. Future costs are not discounted, although the manual mentions that a zero discount rate had been chosen for simplicity. Also, the total costs calculated are not inclusive of sunk costs and cannot be used to calculate a "full cost recovery" fee.

Basis for evaluation: The information provided here is based on a copy of the manual (May, 1992), and on a profile of the tool presented in Weitz et al. (1994).

Contact information: Both versions of the report are available in SRS library (the updated version is recommended).

Westinghouse Savannah River Company Savannah River Site, PO Box 616, Bldg 705-C Aiken, SC 29802
John P. Harley Site Waste Minimization Group

803-557-6332 803-725-3011 (General)

Systems Cost Model (SCM)

Prepared by: Lockheed, Idaho

Prepared for: U.S. DOE Assistant Secretary for Environmental Management (EM-30), under DOE Idaho Operations Office.

Application: financial analysis: - Environmental impact analysis: - Waste management/P2: * Environmental cost listing/database: - Cost estimation: * Alternative product/process comparison: -

Parametric life-cycle cost estimation for radioactive and hazardous waste treatment and disposal. Although developed specifically for the DOE, this model could possibly be applied (with appropriate modifications) to other facilities as well.

Development/publication date and updates: The software has been released, but additional development is continuing. A non- integrated version of SCM called the PEIS model, has been developed. The latest version was brought out in December, 1994. This deals with radioactive waste. Lockheed is currently developing the hazardous waste area of the system.

Public availability: Information on SCM available

Purpose and current use: The system was developed specifically for DOE facilities. SCM will support management planning for the Programmatic Environmental Impact Statement (PEIS) and DOE'S cost baseline for the total program estimate. Initial users of SCM will be the Lockheed Idaho Technical Support Program and MK Environmental. Future users include DOE-HQ EM-33, EM-353,

Boss and Associates, and DOE sites.

Cost information: Free of charge to DOE facilities

Contents and system requirements: Software, user's manual. The system is stand-alone, currently without networking capabilities. The minimum requirements for SCM: « IBM-compatible PC 386 SX (DX recommended)

· Windows 3.1
· DOS 5.0 (or higher)
· 4 MB RAM (8 recommended)
· 10 MB free space on hard disk
· VGA display, and a mouse.

The existing PEIS model requires:

· IBM-compatible PC using DOS
· Lotus 123 Version 3.4
· Baler (of Baler Software Corp.).

Summary of methodology

The non-integrated version of SCM (PEIS model) has the following programming elements.

· Volume Input Model
· Bottom-up model
· Waste Management Estimating Program
· Socio-economic model
· Transportation model

The main task will be to further automate and integrate the PEIS models. Appropriate modifications and developments for SCM are established through prototyping. SCM will not initially include the socioeconomic and bottom-up model (although they may be added later). The minimum features of SCM (Version 1.0) will include the following:

· Static analysis capabilities of the current PEIS model (except the bottom-up and socioeconomic model)
· Integration of remaining three PEIS models
· Expanded user-defined input parameters
· Support of cost and manpower profiles over time
· Expanded Work Breakdown Structure

The system will also include on-line help screens and other features to increase user-friendliness to allow the system to be used by a wider audience. It will also feature a report generator with expanded report generation including graphical output. Output data includes module capacities and costs.

Life-cycle stages covered

Raw materials acquisition
Manufacturing
Use/reuse/maintenance
Recycle/waste management

Type of costs considered

Conventional
Potentially hidden
Contingent
External

SCM includes cost input elements for items such as labor, construction, O&M, D&D, infrastructure, and costs per volume of waste. Users will have the capability to select, enter, or change data including site-specific cost factors, existing or new DOE facility cost information, commercial total or unit costs, off-site DOE treatment total or unit costs, and cost escalation factors. This may be a way to factor in contingent or external costs that have been excluded in the estimation. For example, users can change operating cost contingency percentages to factor in contingent costs due to environmental variables. Cost per unit volume of waste can be modified to include external costs. However, users will need to develop these estimates. No additional cost elements can be added.

Method of cost estimation

SCM uses cost capacity relationships and internal data that have been developed for treatment and disposal processes for radioactive waste. The cost estimating modeling techniques used in the Volume Input Model and the Waste Management Estimating Program Model will be used in SCM. The total life-cycle costs will be calculated in three different modules for the waste types, low level waste, mixed low-level waste, and transuranic waste. SCM will have an eight element WBS:

- Pre-Operations
- Facility Construction Costs
- Operations and Maintenance
- Decontamination and Decommissioning
- Contracted Services
- Off-site Treatment, Storage, or Disposal (DOE)
- Special site costs
- Transportation

SCM estimates costs based on internal cost and schedule data, facility capacity and operating parameters, facility costs, maximum and minimum scaling factors for Parametric cost/capacity equations, transportation miles and costs/mile, standard operating parameters, and a standard rate table. SCM also features **waste** load inventory data from 50 DOE sites including an allowance for complex cost ran-ups.

Thus, users enter input data, or accept default values for a variety of elements including waste load by site and waste type, treatment technology option by site, site schedules by site and waste type, existing or planned approved facilities, and site specific cost factors. SCM will estimate costs based on user-input and default information, and Parametric relationship models.

Generation of financial indicators

Net present value (NPV)
Payback period
Internal rate of return (IRR)
Benefits cost ratio
Others

Attributes

Although the software model is still under development, the methodology can be reviewed in a series of reports that are being prepared by Lockheed that describe preconceptual designs and planning level life-cycle cost estimates for four waste categories: Low level waste, mixed low-level waste, transuranic waste, and hazardous waste.

SCM will be designed to allow easy updating and addition of features. Users can select, enter, or change a variety of data types, which allows the evaluation of various options. Costs of new versus existing facilities can be compared. Anticipated enhancements for versions 1.3 and 2.0 include expanded integration and automation of the cost model and extensions such as risk indicators. Other considerations include interface to an interactive scheduling tool, user specification of module capacities, and variations in mass/volume reduction treatment factors to reflect technology improvements.

Limitations

SCM does not include environmental liability (contingent) and external costs. No financial indicators are generated to compare options and results. However, present value analysis may be included in future versions of SCM. SCM is designed primarily with the purpose of estimating waste management facility costs. It does not assist in generating waste minimization options or pollution prevention.

Basis for evaluation: The profile presented here is based on the report, "Waste Management Facilities Cost Information: System Cost Model Product Description," and through telephone communications with Lockheed in May, 1995.

Contact information: Lockheed Martin, Idaho Technologies
PO Box 1625 Idaho Falls, Idaho 83415-2420
Dave Shropshire 208-526-6800 Mike Sherick 208-526-6922
Jay Rhoderick (DOE, EM-30) 301-903-7211
SCM description:
F. Feizohalli, R.K. Fink, B.L. Peterson, D.E. Shropshire, "Waste Management Facilities Cost Information: System Cost Model Product Description," Lockheed Idaho Technologies Company, November 1994. Report number: INEL-94/0115

The following reports are available in public reading rooms around the country as support documentation for Programmatic Environmental Impact Statement.

1. F. Feizollahi, D.E. Shropshire, "Interim Report: Waste Management Facilities Cost Information for Mixed Low-Level Waste" Report number: INEL-95/0014
Similar reports for other waste types:
2. Low-Level Waste: INEL-95/0013
3. Transuranic Waste: INEL-95/0015
4. Hazardous Waste: INEL-95/0016

Information about the PEIS model:

1. "Waste Management Estimating Program User's Guide," Revision 3, May 1993.
2. "Waste Management Facilities Cost Information Work Plan," EGG-WM-1 1233.

Waste Cost Avoidance Model Template

Prepared by: EG&G (now called Lockheed)

Prepared for: All Department of Energy Facilities

Application: Financial analysis: - Environmental impact analysis: - Waste management/P2: * Environmental cost listing/database: * Cost estimation: - Alternative product/process comparison: -

Framework for accounting for cost elements associated with each waste type and treatability group produced at a site. There is a fundamental basis that describes how costs were collected, that could possibly be applied to other sites and facilities as well.

Development/publication date and updates: Published as part of a final report called "Avoidable Waste Management Cost" (INEL-- 94-0205) in January, 1995.

Public availability: Available

Purpose and current use: Provide basic information and guidelines on methods to calculate waste life-cycle costs. The objective was to support better decision-making concerning waste minimization, generation, facility upgrades, and process changes at DOE facilities.

Cost information: The price of the report through OSTI is $36 (including shipping and handling).

Contents and system requirements: Worksheets

Summary of methodology

This set of worksheets lists input and output costs to be estimated by the user. Work sheets are divided into waste types and treatability groups: high-level waste, transuranic waste, transuranic-mixed, low-level waste, low-level mixed waste, hazardous RCRA, hazardous non-RCRA, TSCA waste, sanitary waste, and other nonregulated wastes.

Life-cycle stages covered

Raw materials acquisition
Manufacturing
Use/reuse/maintenance
Recycle/waste management

The method lists four life-cycle stages (the first tier) of waste management: generation, treatment, storage, and disposal. Each first tier stage is broken down into several inputs and outputs (second tier costs) as follows:

·**processing/manufacturing:** waste generation, characterization
·**packaging/transportation/storage:** packaging, lag storage at each life-cycle stage, in-process transportation, transportation from each life-cycle stage to the next, storage facility replacement costs, emplacement
·**waste management:** treatment: operating costs, materials, disposal facility
replacement cost, handling cost, closure

Type of costs considered

Conventional
Potentially hidden
Contingent
External

The hidden costs include costs associated with the operation of a waste treatment facility, packaging, waste storage, etc. Costs associated with compliance, such as permitting costs, are not included.

Method of cost estimation

The template recommends using the most current (in this case, 1992) waste management costs or older cost data (1991 costs) if the information is deemed better.

Generation of financial indicators

Net present value (NPV)
Payback period
Internal rate of return (IRR)

Benefits cost ratio
Others

Attributes

This method is straightforward and simple to use. Transportation and lag storage between life-cycle stages is considered. There are worksheets for listing definitions and assumptions. The document on "Waste Life-cycle Cost: Generic Guidelines and Assumptions," is recommended as the guide to developing waste life-cycle costs. The fixed and variable costs are also defined in this document.

Limitations

External and contingent costs are not taken into account. Life-cycle costs are computed for the waste management stage of the life-cycle.

Basis for evaluation: The information provided here is based on a profile of the tool presented in Weitz et al. (1994). Subsequent reviews were done by incorporating information obtained through telephone communications with Lockheed and INEL in May, 1995.

Contact information: Available for government agencies and contractors through OSTI. The report will ultimately be sent to the NTIS library (order/accession number: DE-95008669, from where other parties may purchase it).
Lockheed Martin, Idaho Technologies PO Box 1625, Idaho Palls, 1083415
Chris Ischay 208-526-4382
For purchase:
Office of Scientific and Technical Information (OSTI) 615-576-8401
The report is also available at the INEL Technical Library (not for purchase): 208-526-5856
National Technical Information Service (NTIS) 800-553-6847 703-

487-4660

Waste Life-cycle Cost: Generic Guidance and Assumptions

Prepared by: EG&G (now called Lockheed)

Prepared for: All Department of Energy Facilities

Application: Financial analysis: - Environmental impact analysis: - Waste management/P2: * Environmental cost listing/database: * Cost estimation: - Alternative product/process comparison: -

Generic guidance for developing waste life-cycle costs. The guidance is directed toward DOE facilities, and definitions of life-cycle costs are consistent with DOE assumptions. Uses include prioritization of waste minimization activities, cost-benefit analysis of process modifications, and computing current costs of waste processing.

Development/publication date and updates: Published as part of a final report called "Avoidable Waste Management Cost," number INEL-94-0205, in January, 1995.

Public availability: Available

Purpose and current use: The objective was to enhance the ability of various DOE facilities to make better decisions concerning waste minimization, generation, facility upgrades, and process changes.

Cost information: The price of the report through OSTI is $36 (including shipping and handling).

Contents and system requirements: General guidelines, assumptions, suggestions

Summary of methodology

This method considers costs associated with waste materials only. A cost/benefit analysis is recommended for the comparison of options.

Life-cycle stages covered

Raw materials acquisition
Manufacturing
Use/reuse/maintenance
Recycle/waste management

The inputs and outputs considered include the following.

processing/manufacturing: waste generation, accumulation, and characterization
packaging/transportation/storage: waste packaging and handling, transportation capital and operating
waste management: treatment, storage, disposal including allocation of costs for extended storage and disposal capacity), disposal site closure, pre- and post-closure monitoring and surveillance

Type of costs considered

Conventional
Potentially hidden
Contingent
External

Inputs and outputs are divided into generic cost categories:

· **activities associated with fixed costs**

-facility management
-environmental program development
-permitting
-equipment replacements/required upgrades
-quality assurance
-preventive maintenance
-fire protection/emergency management
-records management system-operating procedures development and management training
-generator waste certification program
-site planning and design
-utility construction
-health protection program
-regulatory compliance

· **activities associated with variable costs**
-waste handling/transport
-records origination/management
-earth moving
-spent storage/disposal capacity
-waste packaging
-waste inspection, sampling, characterization
-radiation monitoring
-job control/scheduling
-corrective maintenance
-waste treatment

Method of cost estimation

Costs are divided into three categories: 1. *fixed costs-costs* that remain constant, in total, regardless of business activity,
2. *sunk costs-tests* that have already been incurred, committed, or planned and are not affected by waste generation rates in the present or future (e.g., cost of existing facilities, capital and operating costs

for existing and new facilities required to handle legacy waste regardless of current waste generation), and
3. v*aiable* costs-costs that vary, in total, in direct proportion to changes in the level of activity. Costs are calculated as follows:

Total cost for a waste category = sunk costs for a waste category + fixed costs for a waste category + (sum of unit variable costs for each waste stream within a category) x (forecasted amount of each waste)

Generation of financial indicators

Net present value (NPV)
Payback period
Internal rate of return (IRR)
Benefits cost ratio
Others

A cost/benefit analysis is recommended for comparing options. The guide does not provide information on the calculation of any indicators.

Attributes

Costs for treatment, storage, and disposal facilities are recommended to be consistent. The guidelines recommend the sharing of these cost elements among sites that can be facilitated through the Waste Minimization Subcommittee of the GOGO Waste Management Committee.

Limitations

The guidance is directed toward DOE facilities. Less tangible (internal, potentially hidden) and contingent costs are mentioned, but not included in the costs considered. External costs are not considered.

Basis for evaluation: The information provided here is based on a profile of the tool presented in Weitz et al. (1994). Subsequent reviews were done by incorporating information obtained through telephone communications with Lockheed and INEL in May, 1995.

Contact information: Available for government agencies and contractors through OSTI. The report will ultimately be sent to the NTIS library (order/accession number: DE-95008669, from where other parties may purchase it).
Lockheed, Idaho Technologies PO Box 1625, Idaho Falls, ID 83415-2420
Chris Ischay 208-526-4382
Office of Scientific and Technical Information (OSTI) 615-576-8401
The report is also available at the INEL Technical Library (not for purchase): 208-526-5856
National Technical Information Service (NTIS) 800-553-6847 703-487-4660

WastePlan

Prepared by: Tellus Institute

Prepared for: US Congressional Office of Technology Assessment

Application: Financial analysis: · Environmental impact analysis: - Waste management/P2: * Environmental cost listing/database: - Cost estimation: * Alternative product/process comparison: -
Solid waste planning and management. Applications include policy and aberration analysis.

Development/publication date and updates: 1988. Updates were released in 1990,1992, and 1994.

Public availability: Available

Purpose and current use: The system has been developed as a policy analysis tool. Tellus currently has 36 clients using the system. The software has been used in municipal, state, federal, and international agencies. State waste management agencies have used it for planning and training programs in various states including New York, Indiana, Missouri, Illinois and Minnesota. It has also been applied abroad in countries including China, Mexico, New Zealand, and Canada. There are no users in industry.

Cost information: The cost of the system is $2000 for private sector users, or $1500 for government, non- profit, and academic users.

Contents and system requirements: Software, user-guide. IBM compatible PCs.

Summary of methodology

WastePlan allows users to build integrated solid waste plans. Any number of distinct sectors and waste types can be defined. The generated wastes are sent through user-defined sequences of source reduction, composting, recycling, and garbage collection. Finally, the wastes are routed to various facilities such as transfer stations, incinerators, landfills, recycling, or composting facilities. The results of this simulation describe waste flows, diversion rates, equipment and labor requirements, processing and disposal capacity needs, total costs, and costs per ton. Assumptions regarding technologies, efficiency, etc., can be altered. WastePlan has generally been applied for the following types of analyses.

· solid waste management planning: integrates various aspects of solid waste generation, management, and disposal.
· data organization and management: standard formats for input and output.
· capital and operating cost analysis: least cost system planning for different combinations of options.
· user-fee analysis: modeling option for user-fees including weight and volume-based fees and subscription- based programs.
· sensitivity analysis: analysis of alternative scenarios through changes in assumptions.

Life-cycle stages covered

Raw materials acquisition
Manufacturing
Use/reuse/maintenance
Recycle/waste management

The manufacturing stage is covered to the extent that source reduction is included in the analysis.

Type of costs considered

Conventional
Potentially hidden
Contingent
External

Typical costs included are labor, administrative and operating costs. Some other costs such as long-term care costs for landfills can also be included.

Method of cost estimation

Cost information is estimated and entered into the system by users.

Generation of financial indicators

Net present value (NPV)
Payback period
Internal rate of return (IRR)
Benefits cost ratio
Others

The system calculates the NPV of the solid waste plan.

Attributes

WastePlan features context-sensitive help screens, choice of mouse or keyboard input, and an on-screen display of the entire data structure. Various report and graph options are also available. The system is designed as a very flexible, generic system. WastePlan has an interactive database that takes changes in waste-flow into account. For example, if the generation rate changes, the diversion rates will automatically change.

Limitations

WastePlan does not include contingent, hidden, and external costs in the analysis. It is not designed for life-cycle cost analysis. It focuses on the waste management portion of the life-cycle, and thus does not include other costs and environmental impacts of products, processes, and raw materials that are responsible for generating the waste. The downside of the system flexibility is that there is very little default information in the system, and users have to source and develop a lot of information on their own (or with technical support from Tellus).

Basis for evaluation: The profile presented here is based on information provided by the Tellus lnst. in May, 1995.

Contact information: Multiple and single site licenses are available
Tellus Institute 11 Arlington Street Boston, MA 02116
Paul J. Ligon Email: PLIGON@TELLUS.COM (Tellus Solid Waste Group) 617-266-5400 617-266-8303 (FAX)

REFERENCES

Becker, M., & Ashford, N. (1995). Exploiting opportunities for pollution prevention in EPA regulatory enforcement agreements. Environmental Science and Technology. 29 (5) 220-226.

Chada, N. (1994). Develop multimedia pollution prevention strategies. Chemical Engineering Progress- 90 (11) 32-39.

Chiu, M. F., Shoberg, J. P. (1995). Create a comprehensive chemical information management system Chemical Engineering Progress 91(5) 77- 82.

Davenport, T. H. (1993). Process innovation reengineering work through information technology. Boston, MA: Harvard Business School Press.

Ditz, D., Ranganathan, J., & Banks, D. R. (1995). Green ledgers: Case studies in corporate environmental accounting. Baltimore, MD: World Resource Institute.

Donley, E. M. (1991). Trends in environmental computer applications: Environmental Software 6 (3) 168-169. Donley, E. M. (1993). Trends in environmental computer applications: environmental software mirrors environmental goals:. Environmental Software 8 (3)187- 190. Donley, E. M. (1994). Trends in environmental computer applications: environmental software, more value for less money. Environmental Software 9 *(1)* 67-70.

Environmental Health & Safety Software Development Group. (June, 1995). Functional Requirements Outline. Unpublished document. Various Addresses: Author.

Esry, G.T. (1993). How to automate compliance reporting. Pollution Engineering- 25(9) 40-41.

Federal Register, No. 60, (1995, April 13). Voluntary environmental self-policing and self-disclosure interim policy statement. EPA, 16857.

Federal Register, No. 60, (1995, May 10). Interim revised EPA supplemental environmental projects policy. EPA, 24856.

FitzGerald, C. (1994/1995, Winter). Environmental management information systems: New tools for measuring performance. Total

Quality Environmental Management

General Electric Company (1993). A guide for improving chemical management (second edition). Fairfield, CT. Author.

Guggenheim, D.E. (1994/1995, Winter). An electric utility's challenge: creating an integrated chemical management information system. Total Quality Environmental Management 55-62.

Hammer, M., & Champy, J. (1993). Reengineering the Corporation. New York, NY: Harper Business.

Heilshorn, E. D., & MacDougall, J. J. (1992). Identify and track hazardous wastes effectively. Chemical Engineering Progress, 88 (11) 52-59.

Herman, S. A. (1995). EPA's new voluntary environmental self-policing and self- disclosure interim policy statement. National Environmental Enforcement Journals 5 (5) 15-17.

Hunsley, J. R. (1995). The chemical hygiene plan: PC-based chemical inventory and MSDS's. Journal of Chemical Educations 72 *(6)* 543-544.

Kirschner, E. (1995). Environmental management systems get worldwide benchmark. Chemicals and Environment 13.

Klassen, R. D. (1993). The integration of environmental issues into manufacturing: toward an interactive open-system model. Production and Inventory Management Journals 34(1) 82-88.

Klassen, R. D., & McLaughlin, C. P. (1993). TQM and environmental Excellence in manufacturing. Industrial Management & Data Systems' 93 *(6)*14-22.

Kolluru, R. V. (1995) Minimize EHS risks and improve the bottom line. Chemical Engineering Progress- 91(6) 44-52.

Kumar, A., & Sahore, S. (1993). A review of four software directories. Environmental Progress- 12 (2) M9-M14.

Lorenzi, N. (1994). Striving to achieve environmental, health, and safety excellence. Professional Safety- 39 (8) 34-36.

Luce, Z. R (1990). IMP the integrated management approach to environmental protection, health, and safety. Professional Safety- 35 (1) 30-33.

McAdams, C.L., & McAdams, W. A. (1995). Computer technology in the waste industry. Waste Age- 26 *(9)*125-130.

McLaughlin, M. W. (1995). Software selection demystified. Computerworld, 29 (25) 41.

Nixon, K.(1994). How to choose tile best warehouse control system to meet yd. needs. Material Handling Engineering, 49 (2) 69-72.

Randall, M., & Shih, M. (1990). Benefit at first byte.. Pollution Engineering- 22 *(1)* 70-71.

Rich, G. (1995). Making your environmental software selection. Pollution Engineering, 27 (1) 30-41.

Rich, G. (1992). Rcking environmental software. Pollution Engineering, 24 45-66.

Rich, G. (1995). Environmental Protections annual guide to environmental software. Environmental Protection 29-33.

Rich, G. (1989). Environmental software review: 1989. Pollution Engineering, 21 (1) 80-93..

Rich, G., & Shih, M. (1991). Selecting software for the long-term Pollution Engineering- 23 (1) 56-59.

Shanley, A. (1995) Cleaner by design. Chemical Engineering- 102(8) 32-37. Sheng, P. S., Dornfield, D. A., & Worhach, P. (1995). Integrating issues in"green" design and manufacturing. Manufacturing Review, 8 (2) 23-34.

Sowa, C. J. (1994). Explore waste minirakation via process simulation. Chemical Engineering Progress, 90 (1) 40-42.

U. S. Environmental Protection Agency. (1995) Incorporating environmental cog and considerations into decision-making: Review and available tools and software Unpublished report of the Office of Pollution Prevention and Toxics, Washington, DC: Author.

U. S. Environmental Protection Agency. (May, 1995) Fact sheet on revised interim supplemental environmental projects (SEP) policy Unpublished EPA document prepared by the EPA Office of Enforcement and Compliance Assurance, Washington, DC: Author

U. S. Environmental Protection Agency. (May, 1995) Note to correspondents on SEP policy. Communications, Education, and

Public Affairs (1703) Washington, DC: Author.
U. S. Environmental Protection Agency. (June, 1995) TRI automated form R (AFR) for reporting year 1994 list of approved vendors. Unpublished EPA document prepared by tile Office of Pollution Prevention and Toxics, Washington, DC: Author.
Wu, M. S. (1990). Selecting the right software application package. Journal ofSystems Management 10 *(9)* 28-32.

ADDITIONAL REFERENCES

AACE International. 1992. *Skills and Knowledge of Cost Engineering*. 3rd Ed. Morgantown, WV: AACE Inc.
AACE International. 1995. "Total Cost Management." *Cost Engineering* 37(2):27.
Brown, D.R. 1994. "Levelized Production Cost: An Alternative Form of Discounted Cash Flow Analysis." *Cost Engineering* 36(8):13-16.
Capstone Corporation. August 30, 1994. "Environmental Management Category Report for the Survey of Resources Available for Estimating the Environmental Costs of Major Defense Acquisition Programs." Alexandria, VA: Capstone Corporation.
de la Garza, J.M., and K.G. Rouhana. 1995. "Neural Networks Versus Parameter-Based Applications in Cost Estimating." *Cost Engineering* 37(2):14-18.
Ditz, D., Ranganathan, J., and Banks, Darryl R. May, 1995. Green Ledgers: Case Studies in Corporate Environmental Accounting. World Resources Institute (WRI). Baltimore, MD: WRI Publications.
Dragoo, R.E., and R.A. Letendre. 1994. "Manage Costs in Real Time." *Cost Engineering* 36(2):11-18.
Estimator's Notes. Summer 1993. Composer Gold, Building Systems Design (BSD).
FitzGerald, Chris. 1994/1995. "Environmental Management Information Systems: New Tools for Measuring Performance." *Total Quality Environmental Management* Winter: 21-33.

Hombach, W.G. March 22,1995. "Evaluation of Environmental Management Cost-Estimating Capabilities for Major Defense Acquisition Programs." Communications Training Analysis Corporation (CTAC). Prepared for the Office of the Secretary for Defense (PA&E). Arlington VA: CTAC. Richard H. Rener is the contractor's project director and can be reached at 703-683-4220. William G. Hombach is the subcontractors principal investigator and can be reached at 703-415-7774.

ICF Inc. June 1995. *An introduction to Environmental Accounting as a Management Tool: Key Concepts and Terms,* Prepared for the U.S. Environmental Protection Agency Design for the Environment, Environmental Accounting Project. EPA-742-R-95-001. Washington, DC.

Martin, Bruce. 1992. "Aspects of Cost Control." *Cost Engineering* 34(6):19-23.

McDonald, D.F. Jr. 1995. "Defining and Communicating Total Cost Management Principles in the Workplace." *Cost Engineering* 37*(1)*:20-24.

Owen, J. (Senior Editor). 1995. "Environmental Compliance: Managing the Mandate". *Manufacturing Engineering* 114(3): 59-66.

Parker, D.E. 1994. "Project Budgeting for Buildings." Cost *Engineering 36*(12):13-18.

Ralston, A., and E.D. Reilly (eds). 1993. *Encyclopedia of Computer Science.* 3rdEd. NewYork*NY: VN Reinhold.

Rich, G. 1995. "Software Selection." *Pollution Engineering* January: 31-41.

Schall, L.D., and C.W. Haley. 1986. *An Introduction to Financial Management.* 4th Ed. New York: McGraw Hill.

Schumacher, Lee. 1995. "Quantifying and Apportioning Delay on Construction Projects. " *Cost Engineering 37*(2):11 - 13.

Selg, Richard A. (ed). 1993. *Hazardous Waste Cost Control.* New York, NY: Marcel Dekker Inc.

Society of Environmental Toxicology and Chemistry (SETAC). 1991. "A Technical Framework for Life Cycle Assessment."

January. Washington, DC.

Spinner, M. Pete. 1989. *improving Project Management Skills and Techniques*. Englewood Cliffs, NJ: Prentice Hall.

Spitzer, Martin A., Robert Pojasek, Francis L. Robertaccio, and Judith Nelson. "Accounting and Capital Budgeting for Pollution Prevention." Presented at the Engineering Foundation Conferences "Pollution Prevention-Making It Pay: Creating a Sustainable Corporation for Improving Environmental Quality." San Diego, January 24 - 29,1993.

U.S. Environmental Protection Agency, Office of Communications and Public Affairs. December 1989. *A Glossary of Environmental Terms and Acronym List*. (A-107). Washington, DC.

Watson, Susan. February 1995. "1994 Survey. Personal Computer-based Cost Estimating Software Survey and Software Used at DOE." Prepared for the U.S. Department of Energy, Filed Maintenance. lrving, TX: PEI Corporation.

WEC, Environmental Cost Estimating Task Group. August 1992. *Westinghouse GOCO Environmental Cost Estimating Manual, Cleaning Up the DOE Complex*. Pittsburgh, PA.

Weitz, K.A., J.K. Smith, B.F. Fagg, J.L. Warren (Research Triangle Institute), D. Adolphson, K. Tschritter, and B. Odegard (Sandia National Laboratories). April 1994. *DOE/SNL/EPA Joint Effort to Develop Waste Minimization Techniques for DOE R&D Laboratories*. Prepared for the U.S. DOE, Environmental Restoration and Waste Management, Office of Waste Management, Waste Minimization Division. Research Triangle Park: Research Triangle Institute. The project was managed by Mem Lewis (SNL).

Welcom Software Technology (WST). 1993. *Evaluating Project Management Software*. Houston, TX: WST Corporation.

White, A.L., M. Becker, and J.Goldstein. 1991. *Total Cost Assessment: Accelerating industrial Pollution Prevention through innovative Project Financial Analysis*. Prepared for U.S. Environmental Protection Agency, Office of Pricing Planning and Evaluation, Office of Pollution Prevention. Boston, MA: Tellus

Institute.

Winston, R.E. 1995. "Rapid Method for Capital Investment Decisions." *Cost Engineering* 37*(4):41*-43.

Youngblood, Annette, and Steven Booth. June 1992. *Environmental Remediation Cost and Risk Estimating Software Summary*. Submitted to the U.S. Department of Energy. Los Alamos: Los Alamos National Laboratory.

GLOSSARY

ABC : Activity-based costing. ABC techniques associate cask and profits with products' product families, product mix, and business and manufacturing processes. The important activities in the manufacturing process that consume costs and resources are determined. Cost models then compute the cost of performing each activity and the consumption of activities for each product (Dragoo and Letendre, 1994). ABC techniques can often serve to make many environmental costs, previously hidden in overhead accounts, visible.

Activity : A self-contained unit of work within a project (Welcome Software
Technology, 1993).

Activity base : A measure of volume of business activity that is highly correlated with the amount of cost incurred (Schall and Haley, 1986).

Activity splitting : A method to resolve resource conflicts wherein an activity that has been started is put on hold and another shorter activity utilizes the resource in conflict, after which work on the first activity is completed. Software packages differ in how many activity splits they allow (Welcome Software Technology, 1993).

ADM : See arrow diagram method.

Arrow Diagram : Also known as the critical path method (CPM), ADM is a diagramming Method: (ADM) method used for project planning. It mainly involves drawing an arrow for each job, where the sequence of arrows indicate the flow of work from the beginning to the end of each project. The complete diagram has one beginning and one end point. Arrow junctions are called nodes (or events) (Spiner, 1989).

Back-end costs : Include environmental costs that arise following the useful life of processes, products, systems, or facilities. Examples include D&D and site survey costs (ICF Inc., 1995).

Bar chart : Data can be entered interactively though a barchart screen, which is a graphical representation of a project with horizontal bars on a time scale depicting activity information (Welcome Software Technology, 1993).

Benefit cost ratio: The ratio of total benefits to total costs of a project. A value greater than one implies that a net profit is made. This method recommends using projects that have the highest benefit cost ratio.

Budget estimates : These are prepared with the help of flow sheets' layouts' and equipment details. In other words, enough preliminary engineering has taken place to further define the project scope. An estimate of this type is normally expected to be accurate within +30% or -15%. They are also called "design development," "semi-detailed," "appropriation," or "control" estimates (AACE International, 1992). Also see "order-of-magnitude" and "definitive estimates."

CAA : Clean Air Act.

Capacity restrictions : Scheduling software. Refers to restrictions upon the maximum number of items (activities per project, resources per activity, etc.) that can be handled (Welcome Software Technology, 1993).

Capital budgeting : The process of analyzing alternative investments and deciding whether they will be included in a firm's investment budget (White, Becker, and Goldstein, 1991).

Cash Flow : The actual cash available to a company after expenses and taxes (Winston, 1995).

CER : Cost estimating relationship (Hombach, 1995). See "Parametric cost estimating."

CERCLA : Comprehensive Environmental Response, Compensation and Liability
Act.

Contingency : An amount added to the estimate to allow for changes that experience shows will likely be required. This amount may be derived either through statistical analysis of past project costs or by applying experience gained on similar projects. Contingency usually does not include changes in scope or unforeseeable major events such as strikes or earthquakes (AACE International, 1992).

Contingent costs : Environmental costs that may occur following uncertain future events. These are also referred to as liability costs. Examples include costs of penalties, remediation, and legal expenses (ICF Inc., 1995).

Contingent pricing : Unit prices in the contract, additional to the fixed-price elements, which can be used in lieu of or to supplement the fixed price elements of the contract as the work develops and the actual quantities become known (Selg, 1993).

Contract : Aspects of project management that relate to the daily interaction management: between the owner and contractor. This interaction includes all specific coordination roles between such parties as well as the administration of contract terms and conditions (such as payments, change orders, directions, etc.) (Selg, 1993).

Contract price : Value of the work as established *by* the contract provisions. In lump-sum contracts' the contract price is firmly established. In other types of contracts the contract price is determined by quantity surveys or audits of the contractor's costs (Selg, 1993).

Conventional costs : Costs that are usually addressed in cost accounting and capital budgetings including raw materials' capital goods, and supplies. Some aspects of these costs, such as cost savings through decreased use of raw materials, and utility costs, are sometimes overlooked (ICF Inc., 1995).

Cost engineering : The art and science of total cost management with broad components being cost estimating, cost analysis, cost and schedule control, and cost and value improvement (AACE International, 1992).

Cost estimate : A compilation of all the costs of the elements of a project or effort included within an agreed-upon scope. Estimating is a process accomplished in steps that include the take-off, costing, and pricing. The components of a cost estimate include direct and indirect cost elements, including labor material, equipment (direct types) and taxes, risk, and overheads (indirect types) (AACE International, 1992).

Cost estimating : The determination of resource and material quantities and the predicting or forecasting of associated costs, within a defined scope (AACE International, 1995). It is an essentially computational process that attempts to predict the final cost of a future project, even though all the parameters and conditions are not known when the estimate is prepared. Estimating methods vary considerably, depending on the available information, nature of the project, and the time available to prepare the estimate (de la Garza and Rouhana, 1995). The accuracy of the estimate will vary, depending on the stage of the project and information with which it is prepared.

Cost index : A number that relates the cost of an item at a specific time to the corresponding cost at some arbitrarily specified time in the past. A cost index is useful in taking known past costs and relating them to the present (AACE International, 1992).

Costing : Using the take-off and the information presented in the scope documents to assign cost values to the elements of work previously catalogued (AACE International, 1992).

CPA : Contemporaneous period analysis. Also known as what-if or but-for evaluations. CPA breaks the construction period into discreet time periods and examines the delays as project participants would have (Schumacher, 1995).

CPM : Critical path method. This network technique allows the analysis of the relationship of the timing of the completion of specific work activities (and delay activities) to the completion of the project as a whole (Schumacher, 1995).

Crew : A crew is a group of trade personnel and equipment necessary for a particular task (Estimator's Notes, 1993). A crew database will contain information such as hourly rates for equipment and labor. The estimator can determine unit labor and material costs (per unit of outputs for examples for each glass made) by dividing output (units per hour, for examples glasses made per hour) by crew labor or material costs. With this information, changes in labor or material costs can be incorporated into the estimate (or their effects analyzed) (Schall and Haley, 1986).

Critical path : The sequence of activities through a network with the least float and therefore the longest path (Welcome Software Technology, 1993). The chain of interrelated activities in the schedule network having the longest duration and therefore the earliest possible completion of the project (Schumacher, 1995). Most scheduling software will (at the minimum) compute start and finish dates and critical path.

C/SCS : Cost/schedule control system. C/SCSC: Cost/schedule control system criteria.
Customization: The modification or addition of features by users as per their requirements. This can include adding new fields and generating custom reports (Welcome Software Technology, 1993).

CWA : Clean Water Act.

D&D : Decontamination and Decommissioning (Hombach, 1995).

Definitive estimates : Estimates prepared from very defined engineering data. These include fairly complete plot plans and elevations, piping and instrument diagrams, single-line electrical diagrams, equipment data sheets and quotations, structural designs, etc. This includes all types of estimates, from minimum order-of-magnitude to the maximum definitive types. They are also called "check," "lumpsum," "tender," and "post-contract changed estimates. It is expected to be accurate within +15% or -5% (AACE International, 1992). Also see "order-of-magnitudes and "budget estimates."

Depreciation : The decrease in the value of a capital asset with time. The method of depreciation is set by income tax rules that allow various rates of depreciation. The amount depreciated each year is considered an expense for tax purposes (Winston, 1995)

Direct cost : A cost that can be economically traced to a single cost object.

Discount rate : The time value of money or the rate of interest that a company wants to earn on its investments (Winston, 1995).

DOD : Department of Defense. DOE: Department of Energy.

DOT : Department of Transportation.

EA : Environmental Assessment. EA is a written environmental analysis that is prepared pursuant to the National Environmental Policy Act (NEPA) to determine whether Federal action would significantly affect the environment and thus require preparation of a more detailed Environmental Impact Statement (EIS) (EPA, 1989).

EAC : Estimate at completion.

Earned value : The quantification of the overall progress of a project defined in dollar terms to provide a realistic yardstick for comparison with actual cost to date (Welcome Software Technology, 1993).

Economic and the evaluation of economic considerations and financial requirements of financial analysis : a program, operation, project, process, manufactured product, or service within a specific time frame. It includes the application of fundamental techniques such as the time value of money, rate of return, break-even analysis, and profitability analysis, as well as advanced economic comparison methodologies such as cost optimization studies (AACE International, 1995). ECT Environmental Costing Tool EM Environmental Management (Hombach, 1995).

Emission : Pollution discharged into the atmosphere from smokestacks, other vents, and surface areas of commercial or industrial facilities; from residential chimneys; and from motor vehicle, locomotive, or aircraft exhausts (EPA, 1989).

Environmental cost : The addition of environmental cost information into existing cost accounting: accounting procedures and/or recognizing embedded environmental costs and allocating them to appropriate products and processes (ICF Inc., 1995).

EPA : Environmental Protection Agency. EPCRA: Emergency Planning and Community Right-to-Know Act.

Equipment costs : The cost of equipment that a contractor uses to perform a contract (cranes, bulldozers) and not of the equipment installed permanently as part of the contract. If the equipment is leased, the costs consist of lease costs and the costs of fuel. If the contractor owns the equipment the costs include ownership costs and operating costs (AACE International, 1992).

ER : Environmental restoration. Measures taken to return a site to previolation conditions (EPA, 1989).

Escalation : Provision for an increase in cost of equipment, materials labor, etc., over the costs specified in the contract, due to continuing price-level changes over time. Escalation has the same effect on project costs as interest does on the value of a savings account: each year becomes a new base for calculating escalation for the following year (compounded) (AACE International, 1992).

ES&H : Environmental Safety and Health. ETC: Estimate to complete.

External / societal costs : Costs resulting from impacts on the environment and society for which social costs: firms are not held financially responsible. These include environmental degradation and adverse health impacts for which firms are not held legally liable. These costs are intangible in nature and need to be valued by nontraditional methods (ICF Inc., 1995). Some private costs can also be less tangible (see less tangible costs).

Extremely hazardous substances : Any of 406 chemicals identified by EPA on the basis of toxicity and listed substances: under SARA Title Ill. The list is subject to revision (EPA, 1989).

Facility : Any industrial, commercial, or industrial structure, installation, or building (Selg, 1993).

Financial accounting : Refers to the preparation of financial reports for use by investors, lenders, and others, using generally accepted accounting principles. Environmental accounting in this context refers to the estimation and public reporting of environmental liabilities and financially material environmental costs (ICF Inc., 1995).

FS : Feasibility study. Analysis of the practicability of a proposal (e.g., a description and analysis of the potential cleanup alternatives for a site or alternatives on the National Priorities List). The feasibility study usually recommends selection of a cost-effective alternative. It usually starts as soon as the Remedial Investigation (RI) is underway. Together they are commonly referred to as the RI/FS. The term can apply to a wide variety of proposed corrective or regulatory actions (EPA, 1989).

Float : In CPM scheduling, it is the amount of slack time between a noncritical activity and the critical path. In contract management, the amount of time between the contractor's forecast for completion and the contract times specified in the contract (Selg, 1993).

Float suppression : Reducing the amount of float time in a CPM schedule through clever network techniques such as preferential sequencing, unusual use of lead/lag restraints, and/or inflated durations for activities (Selg, 1993).

Full cost : Industry definition: annual fixed and variable costs for producing a good or service. For pricing purposes, the full cost may also include an allocated portion of selling and administrative costs (Schall and Haley, 1986).

Full cost accounting : Accounting: In the context of management accounting, this is a method of financial and management accounting that allocates all direct and indirect historical costs to a product or process (ICF Inc., 1995).

Environment : Used in the environmental contexts this refers to desirable environmental accounting practices. Full-cost environmental accounting is basically the same concepts except that it aims to highlight the environmental elements (ICF Inc., 1995).

FY : Fiscal year COCO: Government-owned contractor-operated.
Hazardous substance: Any material that poses a threat to human health and/or the environment. Typical hazardous substances are toxic, corrosive, ignitable, explosive, or chemically reactive (EPA, 1989).

Historical costs : Costs that were incurred in a past period.
Indirect cost: A cost that is not directly traceable to the manufactured product is associated with manufacturing two or more units of finished product or is an immaterial cost that cannot be traced to a single unit of finished product (Schall and Haley, 1986). In construction, it is all costs that do not become a final part of the installation. They include, but are not limited to, field administration, direct supervision, capital tools, start-up costs, contractor's fees, insurance, and taxes (AACE International, 1992).

Input controls : Controls needed to verify that the input into the computer process is complete and accurate (Schall and Haley, 1986).

IRR : Internal rate of return. This is the discount rate that results in NPV = 0. The IRR is the rate of interest that will make the present value of the cash proceeds expected from an investment equal to the present value of cash outlays required by the investment. This method does not require any assumptions about reinvestment or discount rates (Winston, 1995). The general rule is to accept projects with an IRR greater than the cost of capital or minimum acceptable rate of return (Brown, 1994).

Labor cost : The base salary, plus all fringe benefit costs and labor burdens associated with labor, that can be definitely assigned to one item of work, product, process area, or cost center (AACE International, 1992).

LCC : Life Cycle Costing. The systematic process of evaluating the life-cycle costs of a products product line, process, system, or facility by identifying life-cycle consequences and assigning measures of monetary values to those consequences (ICF Inc., 1995). LCC is also referred to as Life Cycle Cost Assessment (LCCA). However, the definition of life cycle costing traditionally used by Federal facilities does not typically encompass all life cycle stages or external costs. See the definition of Life cycle assessment for a description of the life cycle stages.

Less-tangible costs : This refers to expenses incurred for corporate image purposes or for maintaining or enhancing relationships with regulators, customers, suppliers, host communities, investors/lenders, and the general public (ICF Inc., 1995). These are private costs, but social and environmental costs can also be less tangible. See "external/societal / social costs."

Life-cycle : A holistic approach to identifying the environmental consequences of a assessment/analysis product, process, or activity through its entire life cycle and to identifying (LCA): opportunities for achieving environmental improvements. EPA has specified four major stages in a life-cycle of a product, process, or activity: raw materials acquisition, manufacturing, consumer use/reuse maintenance, and recycle/waste management. LCA focuses on environmental impacts not costs (ICF Inc., 1995).

Life-cycle design : An approach for designing more ecologically and environmentally sustainable product systems and integrating environmental requirements in the earliest stages of design (ICF Inc., 1995).

Lump-sum contract : A type of contract in which the cost of the work to be performed is shown as an all-inclusive price; also known as a fixed-price contract (Selg, 1993).

Management : The process of identifying, collecting, and analyzing information accounting: primarily for internal purposes (ICF Inc., 1995). Management accounting is meant to support a businesses forward looking management decisions.

Manufacturing costs : All costs from the acquisition of raw materials through production, until the product can be turned over to the marketing division (Schall and Haley, 1986).

Material costs : The costs for everything of a substantial nature that is essential to constructing or operating a facility, both direct and indirect. Manufactured equipment costs generally comprise a large part of material costs (AACE International, 1992).

MSDS : Material safety data sheet. A compilation of information required under the OSHA Communication Standard on the identity of hazardous chemicals, health and physical hazards, exposure limits, and precautions. Section 311 of SARA requires facilities to submit MSDSs under certain circumstances (EPA, 1989).

Multi-project : The ability of a software system to support more than one project at a scheduling: time. However, the level of flexibility and functions performed will vary. For example, some programs allow users only to define subprojects under a master project and may or may not roll back results to the project level (Welcome Software Technology, 1993).

National Priority List : EPA's list of the most serious uncontrolled or abandoned hazardous List (NPL) waste sites identified for possible long-term remedial action under Superfund. EPA is required to update the NPL at least once a year (EPA, 1989).

Network diagram : A graphical representation of a project depicting activity information and the relationships between activities (Welcome Software Technology, 1993). Some programs allow interactive data input and editing in network screens.

Neural network : Computer programs loosely modeled after the biological structure of the human brain; they consist of thousands of interconnected computing units called neurons. An artificial neural network can be constructed to simulate the action of a series of neurons, the analogous word being "processing unit." A neural net is normally constructed by arranging processing units in a number of layers. In its simplest form, a neural net consists of three layers: an input layer, an output layer, and a hidden layer (de la Garza and Rouhana, 1995).

NPDES : National Pollutant Discharge Eliminating System. A provision of the CWA, which prohibits discharge of pollutants into the waters of the United States unless a special permit is issued by EPA, a state, or (where delegated) a tribal government on an Indian reservation (EPA, 1989).

NPV : Net present value. The algebraic sum of the discounted values of cash flows each year during the project (Winston, 1995). The general rule is to adopt all projects with NPVs greater than zero or choosing (between two mutually exclusive alternatives) projects that have the highest NPVs (Brown, 1994).

NTIS : National Technical Information Service

Operational costs : Costs incurred during the operating lives of processes, products, systems, and facilities. Examples include maintenance and energy costs (ICF Inc., 1995).

Order-of-magnitude estimates : Estimates made without detailed engineering data. Examples include an estimates: estimate made from cost capacity curves' an estimate using scale-up or scale-down factors, and an approximate ratio estimate. An estimate of this type would normally be expected to be accurate within +50% or -30%. They are sometimes referred to as conceptual or "ball park" estimates. They are usually prepared using only basic criteria and as desired outputs total square feet, or number of units (AACE International, 1992). Also see "budget estimates" and "definitive estimates."

Organizational : A hierarchical system for organizing pools and resources to provide data breakdown structure roll-up for planning, reporting, and what-if analysis (Welcome Software (OBS): Technology, 1993).

OSTI : Office of Scientific and Technical Information

Overhead/indirect costs : All manufacturing costs other than direct materials and direct labor costs: (Schall and Haley, 1986). A cost or expense inherent in performing an operation that is not traditionally charged to or identified with a part of the work, product, or asset and, therefore, must either be allocated on some arbitrary basis believed to be equitable or handled as a business expense independent of the volume of production (AACE International, 1992).

Parametric cost estimating : This method of cost estimating uses statistical techniques ranging from estimating: simple graphical curve fitting to multiple correlation analysis. In either case, the objective is to find a functional relationship between changes in cost and the factor (or factors) upon which cost depends. Parametric cost-estimating techniques often lead to a mathematically fitted function called a cost-estimating relationship, which is a functional model that mathematically describes the cost of a structure, module, or system as a function of one or more independent variables (de la Garza and Rouhana, 1995).

PA/SI : Preliminary assessment/site inspection. PA is the process of collecting and reviewing available information about a known or suspected waste site or release. SI is the collection of information from a Superfund site to determine the extent and severity of hazards posed by a site. It follows a PA and is more extensive. The purpose is to gather information to score the site, using the Hazard Ranking System and to determine if the site presents an immediate threat that requires prompt removal action (EPA, 1989).

Payback period : The length of time it takes for future net positive cash flows to equal the initial investment. Projects are accepted if the payback period is less than some management guideline. The greatest asset of the payback is its simplicity. However, it ignores all cash flows and the time value of money. On the other hand, the discounted payback reflects the length of time it takes for the discounted future net positive cash flows to equal the initial investment (Brown, 1994).

Performance : The systematic application of methods to monitor performance against measurement: plan and to control cost and schedule as well as changes to scope, configurations and functional/operational requirements. It also aims to measure and determine causes of variance from established baselines so that effective and timely action may be taken (AACE Internationals 1995). The ability to track and calculate cost performance and schedule performance. The former is based upon a comparison of earned value and actual costs and the latter on a comparison of earned value and budgeted costs (Welcome Software Technology, 1993).

Permit : An authorizations license, or equivalent control document issued by EPA or an approved state agency to implement the requirements of an environmental regulation (e.g., a permit to operate a wastewater treatment plant or to operate a facility that may generate harmful emissions) (EPA, 1989).

Project Management : Potentially hidden Costs that are obscured in overhead accounts or overlooked in business costs: decision-makings including costs of upfront, operational, and back-end activities undertaken to comply with environmental laws (ICF Inc., 1995).

Present value (PV) : The value today of money received in the future at a specified discount rate (Winston, 1995).

Pricing : Determining the amount to be charged to the owner/client to fully include the direct and indirect cost items, as well as contingency and profit (AACE International, 1992).

Private costs : Costs that directly affect firm's profits and for which firms are held financially responsible. These are costs that directly affect a firm's bottom line. Private costs are sometimes referred to as internal costs (ICF Inc., 1995).

Production : A department that is directly engaged in a manufacturing activity and department: contributes directly to the content and form of the finished product (Schall and Haley.1986).

Profitability index : This index is the ratio of the present value of after-tax inflows to the (PI): outflows. An index greater than one indicates that a project has a yield greater than the discount rate (Winston, 1995).

Project : A unique goal to be accomplished in a specific time frame, using a limited set of resources and characterized by a sequence of activities (Welcome Software Technology, 1993).

Project management : In business and industry it is defined as the managing and directing of time, materials personnels and costs to complete a project in an orderly and economical manner and meet established objectives of time, cost, and technical results. The three major phases of a project (project phases) are plan, schedule, and control (Spiner, 1989). Our definition of project management (for the purpose of this resource guide) goes beyond the industry definition. In our definition, we include planning, cost estimation, economic and financial analysis, cost engineering, scheduling, and performance measurement.

Quality : A system of procedures, checks, audits, and corrective actions to ensure assurance/quality that all EPA research design and performance, environmental monitoring control: and sampling, and other technical and reporting activities are of the highest achievable quality (EPA, 1989).

Rate of markup : The target rate applied against the cost of a product or service to establish a selling price for goods and services in the external market (Schall and Haley, 1986).
RCRA: Resource Conservation and Recovery Act.

RD: Remedial design. A phase of remedial action that follows

RA/FS and includes developing engineering drawings and specifications for a site cleanup (EPA, 1989).

Regulatory/compliance costs : Costs incurred to comply with Federal, state, or local environmental costs: laws. Examples include costs for notification, waste management, and pollution control (ICF Inc., 1995).

Reimbursable cost : A type of contract where the cost of the work is determined by actual contract: cost incurred by the contractor. Contractors' profit and overhead costs can be a fixed amount or a percentage of the actual costs. Also commonly referred to as a time-and-materials contract (Selg, 1993).

Relational database : A relational database is built and operated in accordance with a relational model of data that provides a simple and intuitive method for defining a database, storing and updating data in it, and submitting queries of arbitrary complexity to it. It provides a consistent foundation for all other topics that database managers commonly cover. It enables individual users to work with customized database schemas instead of all having to use the same underlying schemas (Ralston and Reilly, 1993).

Remedial response : A long-term action that stops or substantially reduces a release or threat of a release of hazardous substances that is serious but not an immediate threat to public health (EPA, 1989). Relevant costs: Future costs that are different under one decision alternative than under another (Schall and Haley, 1986).

Resource : A source of project support assigned to activities (labor, materials, facilities, etc.) (Welcome Software Technology, 1993). RI/FS Remedial Investigation / Feasibility Study : See FS...

Risk : The quantification, normally through statistical methodologies, of the potential known, known-unknown, and unknown variations to the result (McDonald, 1995). Construction: elements of risk include profit and contingency. The amount of profit to be added to factor in risk is very subjective and depends on considerations such as competition, the job market, local conditions, etc. Contingency is the amount added to an estimate to allow for changes or project cost growth that experience shows will likely be required. It can be derived through statistical analysis of past project costs or by applying experience gained on similar projects (AACE International, 1992).

Risk assessment : Environment: the qualitative and quantitative evaluation performed in an effort to define the risk posed to human health and/or the environment by the presence or potential presence and/or use of specific pollutants (EPA, 1989). S

SARA: Superfund Amendments and Reauthorization Act (SARA)

Scheduling: The assignment of the estimated duration and the desired start and finish times to each activity in a project within the overall time cycle required for completion according to plan (AACE International, 1995).

Scope : The equipment and materials to be provided and the work to be done. Scope is documented by the contract parameters for a project to which the company is committed (AACE International, 1992).

Sensitivity/what-if : Examines the effect on the outcome for changes in one or more input analysis.Evaluates various project scenarios by modifying certain data (Welcome Software Technology, 1993).

Service department : A department that provides services or assistance to production departments (Schall and Haley, 1986).

Solid waste Supervised handling of waste materials from their sources through management: recovery processes to disposal (EPA,

1989).

Standard cost : The expected cost of production under normal operating conditions. A standard cost is typically a unit cost concept (Schall and Haley, 1986).

Superfund : The program operated under the legislative authority of CERCLA and SARA that funds and carries out the EPA solid waste emergency and long-term removal activities. These activities include establishing the National Priorities List (NPL), investigating sites for inclusion in the list, determining the priority level on the list, and conducting and/or supervising the ultimately determined cleanup and other remedial actions (EPA, 1989).

Take-off : Measuring and cataloging the quantities of work derived from the scope documents (AACE International, 1992).

Total cost : A hybrid term sometimes used as a synonym for either of the definitions accounting: given to "full cost accounting," or as a synonym for "Total Cost Assessment" (see the definitions of Total Cost Assessment and Full Cost Accounting). The application of professional and technical expertise to plan and management: control resources, costs, profitability, and risk. It represents a systematic approach of managing cask throughout the life cycle of any enterprise, program, facility, project, product, or service (AACE International, 1995).

Total cost assessment : Long-term, comprehensive financial analysis of the full range of internal (i.e., private) costs and savings of an investment (White and Becker).

TRU : Transuranic (Hombach,1995). TSCA: Toxic Substances Control Act.

TSDF : Treatment, storage, and disposal facility. Site where a hazardous substance is treated, stored, or disposed of. TSD facilities are regulated by EPA and States under RCRA (EPA, 1989).

Tracking : The act of progressing the project status, also called updating or statusing (Welcome Software Technology, 1993).

Transferred-in-costs : Accumulated costs of materials, labor and overhead applied in all previous processes or departments (Schall and Haley, 1986).

Unavoidable fixed costs : Costs that cannot be reduced or eliminated by reducing the level of costs: operations or by discontinuing some level of operations (opposite of avoidable fixed costs) (Schall and Haley, 1986).

Unit price contract : A type of contract in which the cost of the work is shown as a schedule of unit prices to be used against a measured set of quantities (Selg, 1993).

UPS : Unit price book

Upfront costs : Include preacquisition and preproduction costs incurred for processes' products, systems' and facilities. Examples include site studies, R&D, and installation (ICF Inc., 1995).

USACE : U.S. Army Corps of Engineers

Value engineering : A technical review process in which changes to the specified facilities are considered in an effort to reduce the total cost of the project. In the case of a contractor-initiated request for value engineering changes, many contracts allow sharing of the recovered savings due to the new idea (Selg, 1993). The objective of value engineering is to develop or design a facility or item that will yield the least life-cycle costs or provide the greatest value while satisfying all performance and other criteria established for it (AACE International, 1992).

Variable costs : Costs that vary in total as the volume of production or sales changes (Schall and Haley, 1986).

VOC : Volatile organic compound

Voluntary costs : Costs that are not required for, or go beyond, compliance with environmental laws. These include among others, costs for community relations, some types of training, audits, and recycling (ICF Inc., 1995).

What-if analysis scheduling : The what-if method uses an anticipated or planned schedule that indicates key milestones, as a baseline to measure project delays (Schumacher, 1995). Also see "Sensitivity/what-if analysis."

Work breakdown : A hierarchical system for organizing projects and activities to provide structure (WBS): data roll-up for planning reporting, and what-if analysis (Welcome Software Technology, 1993).

INDEX

ACGIH **329-332**
APACS Life Cycle Cost Software **397**
ASAPMS Housekeeper **112**
ASEAM **451**, **453**
ASPEN Plus **310**
AutoCAD **98**, **310**
AWARE **39**, **40**, **42**, **43**, **243-249**
CERCLA **174**, **223**, **268**, **287**, **357**, **388**, **497**, **514**
chemical management **2**, **7**, **8**, **10**, **11**, **13**
chemical tracking **2**, **14**, **38**, **274**, **295**, **297**
CHEMTOX **297**, **298**
Clean Air Act **8**, **252**, **268**, **274**, **308**, **315**, **323**, **338**, **344**, **357**, **421**, **496**
Cobra **127**, **133**, **137**, **166-172**
Composer Gold **24**, **28**, **29**, **54**, **55**, **60**, **99**, **101-105**, **107**
computer systems **1**
Corbus Systems **355**
Cost Assignment Method **180**
cost control **19**, **32-34**, **38**, **41**, **43**, **53**, **62**, **69**, **76**, **84**, **93**, **112**, **113**, **124**, **127**, **134**, **139**, **183**, **190**, **193**, **199**, **237**, **242**, **250**, **270**, **318**, **334**
Cost Engineer **113**
COST ESTIMATING **25**, **27**, **28**, **31**, **53**, **57**, **61**, **63-65**, **68**, **73**, **76**, **83**, **84**, **89**, **97**, **110**, **112**, **115**, **116**, **119**, **122**, **123**, **146**, **188**, **193**, **207**, **215**, **220-222**, **224-226**, **228**, **281**, **435**, **471**
cost estimation **19**, **20**, **23**, **24**, **29**, **34**, **36**, **37**, **41-44**, **46**, **51**, **57**, **61**, **66**, **83**, **93**, **108**, **130**, **135**, **152**, **215**, **221**, **227**, **242**, **263**, **266**, **270**, **296**, **304**, **314**, **321**, **334**, **346**, **354**, **359**, **373**, **378**, **389**, **399**, **405**, **413**, **425**, **430**, **440**, **463**, **466-468**, **482**, **511**
Cost Time Management **32**, **127**, **172**
Cost tracking **20**, **37**, **39**, **40**, **42**, **43**, **236-238**, **240**, **241**, **243**, **244**, **263**, **272**, **281**, **318**, **338**
COSTPRO **23**, **61-67**
cradle to grave **3**, **243**, **282**, **285**, **287**

CSI format **119**
DataPipe **39**, **243**, **250**, **252**, **253**, **255**-**257**
Docuwaste **243**, **291**-**297**
DOE Facilities **459**, **468**, **475**, **478**, **481**
DOT **1**, **15**
EasyABC Plus **33**, **34**, **127**, **177**, **178**, **180**, **182**
ECO-SYS **401**, **402**, **405**, **407**
EE Energy **408**
Emission Reduction Research Center **309**
ENFLEX Data **243**, **299**, **300**, **302**
Enhanced Cost Estimating Relationship Program **115**
Environmental Health & Safety **487**
EOS **28**, **84**-**91**, **144**, **146**, **147**, **188**
EPCRA **1**, **15**
ERMIS **34**, **35**, **193**, **211**, **212**
Excel **31**, **170**, **173**, **266**, **270**, **340**, **341**, **379**, **397**, **399**, **400**, **464**
Facility Pollution Prevention Guide **50**, **51**
FAST **24**, **29**, **68**-**72**, **74**, **75**
FLOW GEMINI Information System **243**, **272**
FOREFRONT **127**, **188**, **189**
Form R **13**, **252**, **260**, **261**, **268**, **274**, **294**, **301**, **315**, **317**, **332**, **338**, **345**, **357**
G2 Estimator **76**, **81**, **82**
G2 Real Time Expert System **39**, **41**, **243**, **305**, **308**, **309**
Gentrax **39**, **243**, **313**-**319**
GEOPHYSICS **365**
GIS capability **316**
Hard Dollar **28**, **84-86**, **89-91**, **143**, **144**, **146**, **147**, **188**
Hard Dollar EOS **28**, **84-86**, **144**, **146**, **147**, **188**
hazardous waste **54**, **56**, **58**, **174**, **194**, **216**, **226**, **261**, **263**, **394**, **421**, **422**, **459**, **466**, **468**, **472**, **474**
Hazardous Waste Manifests **351**, **354**, **358**, **365**, **376**
HAZE **243**, **321**

HAZMAT **48**, **49**, **417**-**419**, **422**-**424**
HAZMIN **43**, **243**, **328**-**332**, **335**
HCAS **51**, **109**, **110**, **189**
High-end scheduling **19**, **127**
ICECAN **35**, **193**, **198**
Industrial Waste Prevention **50**, **51**
LCA Matrix **449**, **451**, **453**
life cycle **23**, **24**, **37**, **40**, **45**, **48**, **51**, **54**, **56**, **95**, **102**, **135**, **140**, **145**, **158**, **164**, **169**, **174**, **186**, **195**, **204**, **209**, **246**, **295**, **311**, **324**, **359**, **397**, **398**, **428**, **429**, **431**, **432**, **435**, **436**, **440**-**442**, **446**, **451**, **452**
Lotus 123 **81**, **170**, **380**, **469**
Low-end scheduling **19**, **32**, **127**, **161**
Low-high end scheduling **19**, **127**, **156**
Macintosh **113**-**117**, **120**, **178**, **182**, **231**, **262**, **315**, **363**, **364**, **451**, **464**
mapping **7**, **260**, **316**
Material Inventory Report System **39**, **243**, **336**, **338**
Material Safety Data Sheet **259**, **294**, **331**, **506**
Monte Carlo **20**, **34**, **35**, **134**
Monte Carlo simulation **194**, **197**, **203**, **208**, **210**
MOUSE **35**, **193**-**198**
MSDS SOLUTION **329**-**332**
neural networks **309**
NPDES permit **345**
Open Plan Pro **127**, **132**-**137**, **162**, **166**, **167**, **170**
OSHA **6**, **15**, **65**, **79**, **259**-**261**, **268**, **275**, **294**, **323**, **330**-**332**, **339**, **357**, **358**, **365**, **388**, **506**
Output Measure Method **179**
PACES **117**, **118**
PAPA Investment Model **394**
Parade **127**, **139**, **141**, **183**-**187**
Pareto's law **208**, **210**
PERT **113**, **157**
Plantware **39**, **40**, **243**, **342**-**346**, **348**

pollution prevention **1-4**, **9**, **16**, **45**, **50**, **245**, **249**, **280**, **281**, **291**, **327**, **338**, **369**, **373**, **377**, **378**, **385**, **389**, **394**, **406**, **422**, **459**, **462**, **473**, **489**, **492**
PRECOSIS **45**, **46**, **389**, **391-393**
Primavera Project Planner **138**, **203**, **204**
Project Management System **112**, **119**, **127**, **153**, **162**, **189**
project planning **33**, **34**, **43**, **134**, **148**, **157**, **160**, **243**, **249**, **266**, **271**, **279**, **285**, **313**, **345**, **495**
project scheduling **19**, **112**, **134**, **139**, **236**
Pulsar **118**
Questimate **120**
RACE **122**
Range Estimating **35**, **57**, **206**, **208**
RCRA **1**, **3**, **5**, **15**, **174**, **223**, **268**, **275**, **287**, **294**, **317**, **339**, **363**, **388**, **475**, **511**, **515**
remediation **20**, **37**, **38**, **108**, **124**, **189**, **190**, **212**, **215**, **219**, **220**, **222**, **226**, **233**, **235**, **239**, **269**
remediation technologies **232**, **233**
Risk assessment **19**, **34**, **36**, **193**, **205**, **206**, **211**, **234**, **260**, **279**, **281**, **283**, **302**, **513**
SARA **6**, **10**, **12**, **252**, **260**, **268**, **274**, **292-294**, **297**, **300**, **301**, **315-317**, **323**, **330**, **332**, **338**, **345**, **357**, **360**, **503**, **506**, **514**
SRS **463**, **466**, **467**
StreetSmarts **123**
Systems Cost Model **468**
TEAM **48**, **49**, **115**, **440-444**
Tellus Packaging Study **48**, **444**
Texim Project **127**, **133**, **161-164**, **166**, **168**
TINIA **40**, **42**, **43**, **243**, **278-284**
TRACES **101**, **102**, **105-109**, **111**
TSCA **8**, **252**, **268**, **301**, **308**, **315**, **323**, **330**, **338**, **475**, **514**
VOC **8**, **339**, **340**, **364**, **365**
waste life-cycle cost **475**, **477**, **478**

WASTE REDUCTION SYSTEMS **50**, **459**
waste tracking **10**, **38**, **245**, **267**, **285**, **287**, **295**, **301**, **339**, **358**
Waste Trail **243**, **355**-**360**
WastePlan **482**, **483**, **485**
Windows **28**, **54**, **62**, **70**, **76**, **83**, **98**, **100**, **101**, **104**, **105**, **114**, **119**, **120**, **125**, **127**, **128**, **133**, **134**, **136**, **138**, **139**, **141**, **147**, **149**, **157**, **161**-**163**, **165**, **172**, **173**, **178**, **182**, **185**, **189**, **190**, **203**, **204**, **211**, **216**, **221**, **231**, **241**, **251**, **255**, **259**, **261**, **277**, **286**, **287**, **290**, **292**, **293**, **299**, **300**, **312**, **314**, **322**, **327**, **341**, **343**, **344**, **353**, **356**, **379**, **397**, **402**, **441**, **444**, **456**, **469**
WinEst **124**
Wixel products **39**, **313**
Woodward Clyde **34**, **35**, **211**-**213**